AF531632

UNDERSTANDING BIOGEOGRAPHY

UNDERSTANDING BIOGEOGRAPHY

By

Dr. Veena
Dept. of Zoology
M.M.H. College
Ghaziabad (U.P.)
(India)

DISCOVERY PUBLISHING HOUSE PVT. LTD.
NEW DELHI-110 002

First Published - 2010

Reprinted - 2017

ISBN: 978-81-8356-507-3

© Author

Understanding Biogeography

Published by:

DISCOVERY PUBLISHING HOUSE PVT. LTD.
4383/4B, Ansari Road, Darya Ganj
New Delhi-110 002 (India)
Phone: +91-11-23279245, 43596064-65
Fax: +91-11-23253475
E-mail: discoverypublishinghouse@gmail.com
sales@discoverypublishinggroup.com
web: www.discoverypublishinggroup.com

Printed at:
Infinity Imaging Systems
Delhi

Preface

The present title "Understanding Biogeography" has been written for those students interested in careers in diverse fields of biological sciences. It provides a structured approach to learning by covering all the important topics in a uniform, systematic format. The book has been comprehensively designed incorporating recent advances in this fast moving field. It also provides accessible information on biogeography in compact form for undergraduate students in biology and related life sciences. It is intelligible to the educated layman, though it deals with some complex ideas. It is an adequate text for all the requirements of students in this area. In addition, busy lecturers who require a quick reference compendium will find it useful, particularly for tutional planning. Simple, yet hopefully clear figures and tables are provided throughout the book.

The over-riding goal of this book, and indeed of the whole *Understanding series*, is to present the essential information concering biogeography in a compact, readily accessible form which leads itself to student learning and revision. The convergence of various approaches has generated a rich panorama of detail, the significance of which we are still attempting to unraval. The present text has been written as an introduction to this rapidly growing field.

To make the work more comprehensive and informative, the author has consulted many authoritative books, research journals, abstracts, monographs etc., so there can be no claim to originality except in the manner of treatment.

The author expresses his thanks to his friends and colleagues whose continue inspirations have initiated him to bring out this book.

The author expresses his gratitude to Mr. Wasan and staff of M/s Discovery Publishing House Pvt. Ltd. for their whole hearted co-operation in the publication of this book.

In the mean time, the author will remain sincerely responsible for any shortcomings of the book and be grateful to the readers for their suggestions and constructive criticism for the continuous betterment of the book. He takes this opportunity to appeal to the readers to send their suggestions straightaway to his Publisher.

Author

Contents

1

INTRODUCTION

In the other chapter of this book the evolution of the Universe from the Big Bang to its present state, with particular attention to those events that were most *crucial* for the eventual genesis of life on Earth. Following the Big Bang, the primordial hydrogen and *helium* condensed into stars, *galaxies*, and galaxy *clusters*. Within the cores of stars the hydrogen was fused into helium, carbon, nitrogen, oxygen, and other heavier elements.

At the end of their lives, the most massive stars exploded as supernovae and threw the newly synthesized elements into space, thus altering the composition of the *interstellar* gas and dust within *galaxies*. Finally, some 4.5 billion years ago, our Solar System was created in the disk of the Milky Way.

The third of the Sun's planets, Earth, was rather special. It satisfied all the conditions required for life to begin and to evolve. It possessed ample chemical raw materials of the right kind, a solid surface, and the correct mass to retain an atmosphere of moderate depth. Its distance from the Sun was such as to produce a climate that was and still is quite mild.

Over large regions of the Earth the temperature was slightly above the freezing point of water, allowing the development of conditions suitable for the occurrence of a great variety of chemical reactions. Equally important, the Sun was born as a star of modest mass, thereby offering for billions of years a steady source of radiant energy, characterised by moderate intensity and a spectral distribution that peaks in the visual range.

Up to the time of the formation of the Earth, the evolution of the

Universe was strictly physical. (Disregard for now the possibility that life-producing planets formed earlier elsewhere in our or in other galaxies, a topic discussed in the Epilogue.) This evolution was based on the existence of five elementary particles (electrons, protons, neutrons, neutrinos, and photons) and their interactions by four fundamental forces (gravitational, weak, electromagnetic, and nuclear forces).

Furthermore, it was based on the presence of plenty of energy, which came into existence in various forms along with the particles during the Big Bang. In principle, physical evolution was and continues to be relatively simple and, in a statistical sense, predictable: Given the same starting conditions, physical evolution will follow a similar path every time.

For example, should the Universe start over again, with the same initial conditions, the end result would again be stars, galaxies, and galaxy clusters, although they and their distribution would not necessarily be identical in every detail to those in our Universe.

With the origin of life on Earth an entirely new kind of evolution beganbiological evolution. Biological evolution is consistent with all the laws of physics, but it is far more complex than physical evolution and much less predictable. This is an important statement and clearly requires justification.

I shall present my justification in the next two sections. In the remainder of this introduction we shall discuss some of the characteristics of organisms, introduce concepts from organic chemistry needed in the chapters to come, describe the genetic code, and conclude with a brief history of the theory of biological evolution.

THE PHYSICAL BASIS OF BIOLOGICAL EVOLUTION

Biological evolution is consistent with the laws of physics because living organisms are made of ordinary matter. This matter consists of organic molecules, water, ions, and trace elements. The organic molecules are composed mainly of atoms of hydrogen, carbon, nitrogen, oxygen, phosphorus, and sulfur. Water molecules are composed of atoms of hydrogen and oxygen.

The ions and trace elements include calcium, potassium, sodium, chlorine, iron, zinc, and others. The atoms of these materials are no different from those found in rocks, in stars, or in interstellar space; consequently, they are subject to the laws of physics, just as atoms are anywhere else in the Universe.

Even though organisms are subject to the laws of physics, they do distinguish themselves from nonliving objects. They possess unique structural organisations that give rise to a range of controlled functional processes geared toward their survival and reproduction. These functional processes give organisms the quality we call living.

Table 1.1 : Major Chemical Elements Present in Living Organisms.

Major constituents of organic molecules (99.5% of all atoms)		*Ions (0.5% of all atoms)*		*Trace elements (less than 0.01% of all atoms)*	
H	Hydrogen	Ca^{++}	Calcium	Fe	Iron
O	Oxygen	K^{+}	Potassium	I	Iodine
C	Carbon	Cl^{-}	Chlorine	Cu	Copper
N	Nitrogen	Na^{+}	Sodium	Zn	Zinc
P	Phosphorus	Mg^{++}	Magnesium	Mn	Manganese
S	Sulfur			Co	Cobalt
				Cr	Chromium
				Se	Selenium
				Mo	Molybdenum
				F	Fluorine
				Sn	Tin
				Si	Silicon
				V	Vanadium

In most organisms the functional processes are enormously complex as, for example, the metabolic activities in our cells, our sensory awareness of external and internal stimuli, the neuronal signals in our brains and the rest of our nervous systems, our bodily movements, and so forth. However, as discussed in other chapters, such extreme functional complexity is not a prerequisite for being alive.

Compared to the processes in our bodies and cells, those of some unicellular organisms appear quite simple, though they are still far more complex than most interactions among nonliving things. In addition, all life is part of an ecosystem that contains many components, interacting with each other in a multitude of ways.

The fundamental unit of biological organisation is the cell. Cells are small, membrane-enclosed bodies, filled with a concentrated aqueous solution of chemicals, and containing all the biochemical machineries required for their metabolism and reproduction. Some organisms -the bacteria and protistsexist mainly as single cells, though a few form transient multicellular assemblages. Other organisms - the fungi, plants, and animals - consist of large numbers of cells precisely organised into specialised, hierarchical structures.

In the case of the animals, these structures include tissues, organs, and organ systems. For instance, in our bodies, the circulatory system is composed of such organs as the heart, the blood vessels, and the lungs; the organs consist of various kinds of tissues, including muscles, membranes, and nerves; and the tissues are constructed from individual cells and their extracellular products.

The range of functional processes in our bodies includes the pumping of the heart and flow of blood through arteries and veins, exchange of oxygen and carbon dioxide across cell membranes and across the delicate linings of the alveoli of our lungs, ingestion and digestion of food, propagation of nerve impulses, movement of arms and legs, conduction of heat, and many more.

Identical or nearly identical processes occur in other animals and similar ones are present in plants, fungi, and single-celled organisms. Ultimately, these processes have their origin in chemical reactions, which are based on interactions between atoms via the electromagnetic force. Again, this means that these processes are in principle no different from those found in the inanimate world and, like them, are subject to the laws of physics. It does not matter that they happen to take place in living things.

THE GENETIC BASIS OF BIOLOGICAL EVOLUTION

The unique structural organisation of organisms and the accompanying functional processes are the result of biological evolution. This evolution distinguishes itself from strictly physical evolution in that it is based on the existence of genetic information and on the changes this information undergoes with time.

Genetic information is the blueprint by which organisms reproduce, grow, and maintain themselves. It is present in nearly all cells, the only exceptions being certain nonreproducing cells such as the red blood cells of mammals. With the exception of red blood cells, every one of the 10^{14} cells in our bodies carries a complete set of genetic information.

In today's organisms, genetic information is encoded in long molecular strands called *deoxyribonucleic acids* or *DNA*. Molecules of DNA are constructed from atoms of carbon, hydrogen, nitrogen, oxygen, and phosphorus, and like the rest of the materials of organisms, this construction is based on physical laws. As discussed in the last section of this introduction, it is the way DNA is put together that allows it to carry information.

During each generation, DNA is replicated and passed on to the succeeding generation. Although this happens with great fidelity, DNA is subject to mutations and other changes that alter its informational content. In the course of many generations, these changes accumulate, giving species the remarkable ability to adapt to changes in their environment and to evolve.

In contrast, objects created without the benefit of genetic information, such as raindrops, rocks, stars, and galaxies, possess no capacity for adaptation to their environments and they do not evolve in a biological sense.

Chance and Necessity

How did genetic information originate and how does it change with time? These two questions are central to any discussion of the origin and evolution of life. Both of them will be examined in the following chapters. Here we shall give just a brief answer to the second one.

The changes in genetic information come about in a two-step process of chance and necessity. In the first step, *random* mutations and their recombinations (if reproduction is sexual) alter the genetic information of a species from generation to generation, thereby contributing to and maintaining a large degree of genetic variation among the individual members of a population.

This variation is expressed by differences both in the outward appearances of the members of a species and in their internal chemistries. We can readily observe this in ourselves, for each of us is a unique individual different from everybody else. We all look different and our bodies react in their own particular ways to heat, cold, food, drugs, disease, and other stimuli.

To be more specific, the number of different ways genetic information can be expressed in human eggs and sperms, for example, has been estimated to be approximately 10^{2000} This is a staggeringly large number and quite beyond our comprehension. It exceeds manyfold the total number of protons present in the Universe, which is a mere 10^{80}. Of course, only a minute fraction of those 10^{2000} possible sperms and eggs will ever be produced.

Therefore, it is most improbable that any two will be identical. This means that every one of the 5 billion people inhabiting the Earth today is genetically distinct and different from everyone else, with the exception of multiple births from the same fertilised egg. Thus, the genetic variations of our species, like that of all other species, are enormously large.

The second step in biological evolution results from the fact that some members of every species are, because of their particular genetic makeup, more competent than others in the acquisition of food and shelter, defense against predators, resistance to disease, and adaptation to changes in climate and other physical conditions.

Hence, some members are more likely than others to survive, reproduce, and pass their genetic information on to the next generation. This inequality in the likelihood of passing on genetic information constitutes a selection process. It continually pushes the genetic makeup of a species in directions dictated by the *necessities* of adapting to the biological and physical environments. The result is biological evolution.

It must be emphasized that the first step in biological evolution is strictly *random*. Mutations and other changes in genetic information occur according to chance, without regard to any benefit or handicap for the individual or the species. In contrast, the second step is imposed by the requirements of the environment and, hence, is *specific* and nonrandom.

It acts like a filter on the reservoir of genetic variation created by the first step, mercilessly rejecting individuals with inadequate competence in the struggle for survival and passing on those best adapted to the current conditions. If the variations in a species do not meet the demands of the environment, its members are filtered out and the species becomes extinct. If they do meet the demands, the species survives and evolves.

The two-step theory of biological evolution was developed during the first half of the 19th century by two British naturalists, Charles R. Darwin and Alfred R. Wallace. Today we often refer to it as "Darwin's theory of evolution." During the century and a half since its inception, the theory has been extended, refined, and thoroughly tested by laboratory experiments. In its broad outlines, it is accepted today by virtually all scientists. However, there continues to be heated opposition, mainly from adherents of fundamentalist religions, just as there has been in the past.

Ecosystems

The full complexity of life and its evolution can be appreciated only if we keep in mind that each species is part of the environment in which it lives. This environment is the totality of its biological and physical components. Each species interacts with and influences its environment and, in turn, is influenced by it. Hence, biological evolution does not occur in isolation. It is a multidimensional process containing countless feedback loops, with all the species evolving together and affecting each other's environment as they do so. The environment—including its biological and physical components—is commonly referred to as an *ecosystem.*

An ecosystem is never static, but is always changing or evolving. The rate of this evolution may vary greatly with time. In general, the evolution of an ecosystem depends on many factors, such as changes in climate or geology, the extinction of certain species, or the development of new adaptations in still other species. Sometimes these changes are sudden and severe, and new ecological niches arise rapidly.

As discussed in later chapters, biological evolution progresses then in an explosive fashion. Species possessing sufficient genetic variation radiate into the newly created ecological niches, like air rushing into a vacuum, and evolution leads to new and distinct species within a relatively short span of time.

The combined effects of production of genetic variation and of selection according to the demands of the environment have created the multitude and diversity of life all around us today and of which we are part. This evolution began more than 4 billion years ago with random chemical reactions in the ocean and atmosphere of the primitive Earth.

Countless generations of organisms have since come and gone. Each represented a temporary link in time. Species evolved and adapted to the ever-changing challenges of the environment. Many succumbed and died out, others prospered and gave rise to new species. In the course of time, biological competition and complexity increased.

Life invaded the land and the air. We and the rest of present-day life are the products of this age-old struggle for survival, and in our DNA we carry the information accumulated by trial and error and by the persistent pressures of natural selection during all those years. This information is our most precious possession.

Unpredictability in Biological Evolution

Biological evolution is not only highly complex, but to a large degree it is also unpredictable. For instance, we cannot be certain what the environment will be like in the future. Hence, we cannot predict which selection pressures will be at work.

Furthermore, we do not understand biological adaptation well enough to know which of several alternatives will do best in a given environment. Once a particular adaptive strategy is adopted, it will be refined by natural selection, but we cannot tell how this will be done. Even if we had perfect knowledge of the evolution of the physical conditions of our planet and thoroughly understood biological adaptation, we still could not predict precisely how life would evolve.

The reason is that the number of different ways genetic information can be written on DNA molecules is so large—10^{2000} different ways in the case of human eggs and sperm—that the possibilities can never all be tried out by living organisms, even in a time span many times the present age of the Universe.

Because mutation and the shuffling of genetic information by sexual reproduction are random processes, we can never know which of the many possible genetic variations will be present in a species. Hence, we cannot predict what the result of selection will be.

Of course, the constraints imposed by the physical environment force any evolving life into certain specific channels. For instance, there is little doubt that life on a planet like Earth would always be based on the chemistry of carbon, hydrogen, nitrogen, and oxygen; that it would start in the sea and eventually spread onto land and into the air; and that photosynthesis and respiration would become the major energy-producing mechanisms.

But there is no way of predicting that insects would become the most populous class of animals, that after 3 or 4 billion years of evolution placental mammals would roam on land, and that a species of primates would evolve with the ability to think, to reason, and to develop culture.

TWO KINDS OF CELLS: EUKARYOTES AND PROKARYOTES

Traditionally, the various life forms that evolved on Earth have always been grouped into two kingdoms—plants and animals. Organisms that are rooted in the ground and obtain their food by photosynthesis were called plants, while those that move about and obtain their food by eating were called animals.

However, after the discovery of microorganisms early in the eighteenth century, it became clear that some of them do not readily fit into this simple classification scheme. For example, euglenas are microscopic aquatic organisms that swim and swallow solid food. Yet they are also capable of photosynthesis. Are they animals or are they plants?

In order to deal with such problematic organisms, one of Darwin's disciples, the German zoologist Ernst H. Haeckel, introduced a third kingdom, the protists. It included protozoa, algae, fungi, and bacteria. But this still did not solve all of the taxonomic difficulties. Numerous other, more complex classification systems were, therefore, proposed.

The one most widely accepted today is the five-kingdom system introduced in 1959 by the Cornell biologist Robert H. Whittaker, which consists of the monera, protists, fungi, plants, and animals. Examples of these five kingdoms, along with summaries of their most important character-istics, are given in figures elsewhere in this chapter.

In addition and complementary to the five kingdoms, biologists recognize today two superkingdoms—the eukaryotes and prokaryotes (terms that mean "true nuclei" and "prenuclear," respectively). The prokaryotes comprise the members of the monera, namely the bacteria.

Their DNA exists freely in the cell interior and is not surrounded by a nuclear membrane. The eukaryotes comprise the members of the other four kingdoms. Their DNA is usually contained within a membrane-enclosed body—the *nucleus*.

The eukaryotes and prokaryotes represent two distinct kinds of cells whose existence has been known only since the early 1950s, when the development of the electron microscope permitted detailed study of cellular architecture.

It became apparent that differences between the two kinds of cells in reproduction, conversion of genetic information into protein structure, and metabolism are related to differences in their internal organisation. These differences were recognised as constituting the most profound demarcation in the contemporary biological world. They are more profound than the differences between us and trees, fishes and mushrooms, or insects and algae.

Eukaryotes

Most eukaryotic cells have sizes in the range between 5 and 60 μm, but some highly specialised cells, such as our nerve cells, may run the length of an entire arm; the largest single eukaryotic cells, such as the giant amoeba Chaos chaos, reach diameters of up to a few centimeters.

All these cells are bounded by thin membranes through which they absorb nutrients from the environment and expel waste as well as products they manufacture, such as proteins, fats, and hormones. Their interiors are filled with a dynamic, aqueous solution called *cytoplasm,* which contains the various raw materials required for growth, self-maintenance, and reproduction: amino acids, fatty acids, sugars, minerals, and many more.

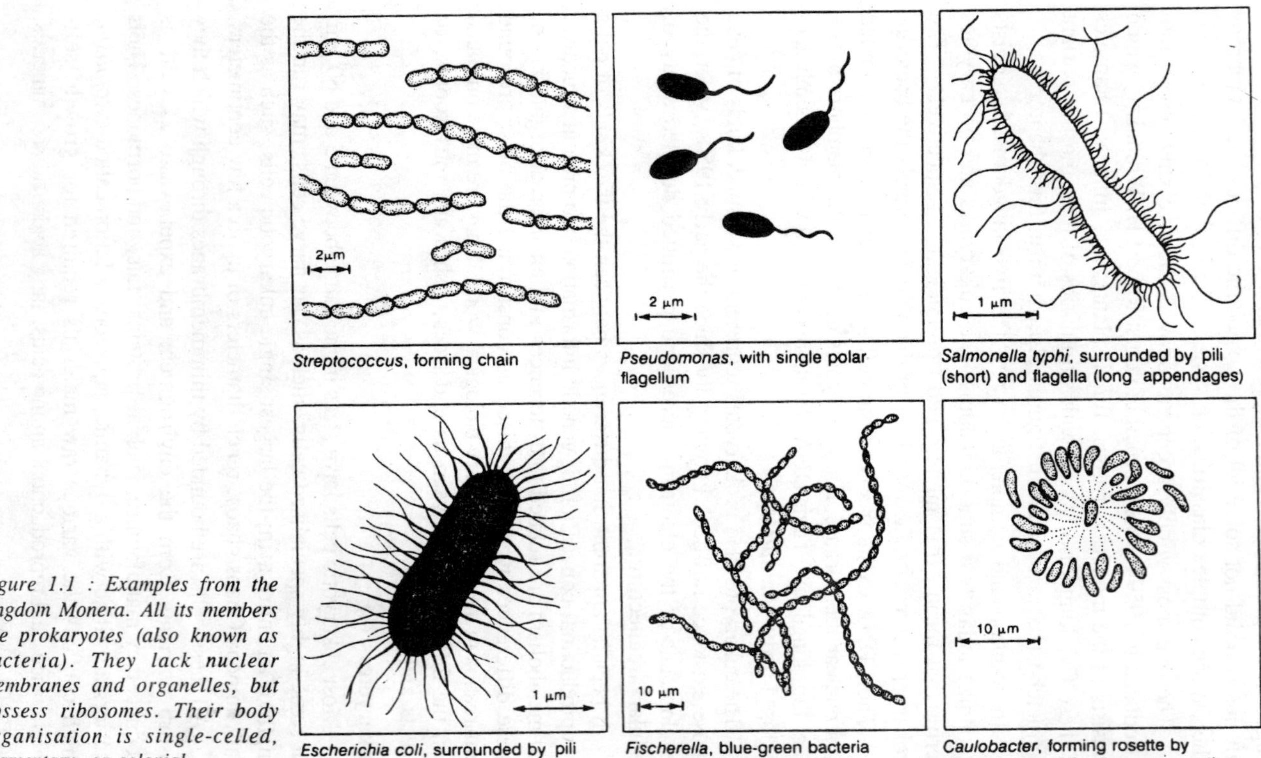

Figure 1.1 : Examples from the kingdom Monera. All its members are prokaryotes (also known as bacteria). They lack nuclear membranes and organelles, but possess ribosomes. Their body organisation is single-celled, filamentary, or colonial.

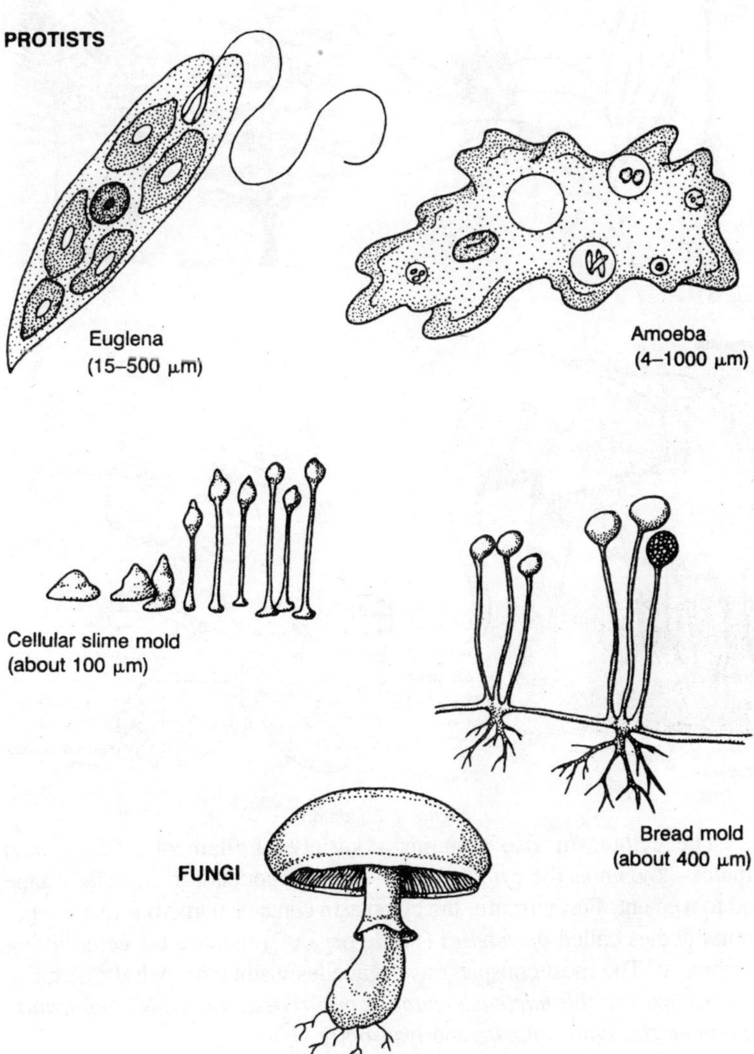

Figure 1.2 : Examples from the four kingdoms of eukaryotes : Kingdom Protista, Kingdom Fungi, Kingdom Plantae, Kingdom animal

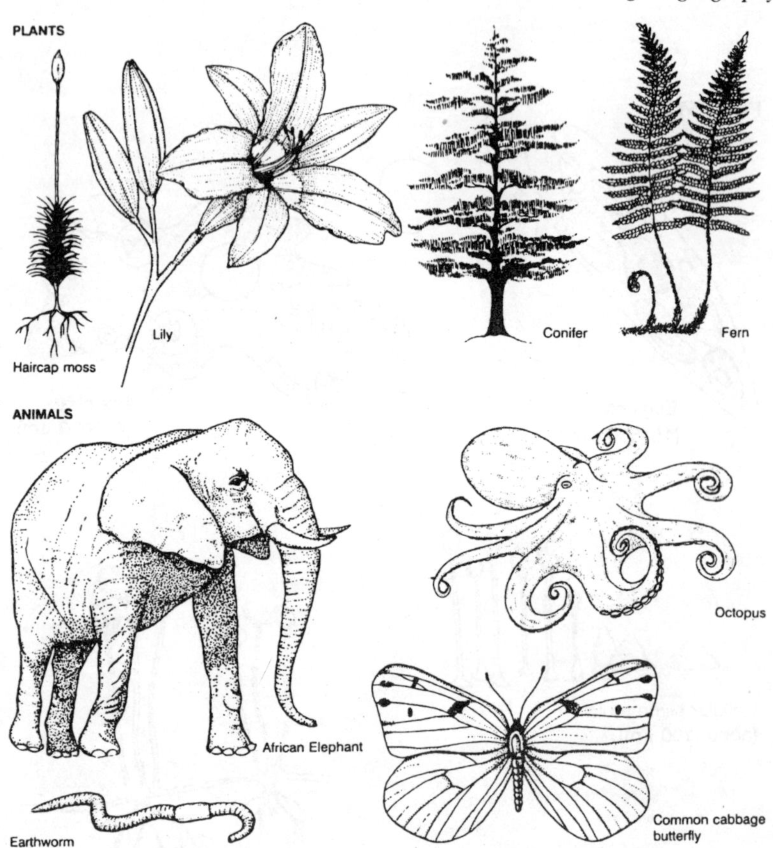

Figure 1.2 (contd.)

The cytoplasm also contains a variety of filaments, fibers, and tubules—known as the *cytoskeleton*—which accounts for the cells' shape and movement. Furthermore, the cytoplasm contains numerous membrane-bound bodies called *organelles* ("little organs") that are the cells' living machinery. The most conspicuous organelles visible through the electron microscope are the *nucleus*, *endoplasmic reticulum*, *Golgi apparatus*, *mitochondria, chloroplasts,* and *lysosomes*.

The chloroplasts are present only in cells that carry out photosynthesis, such as the cells of green plants and algae. Each of the organelles is designed to perform specific tasks, much like the tools and machines in a workshop are designed for specific tasks. Other bodies present in the cytoplasm are the *ribosomes,* which are complexes of protein and nucleic acid and constitute the factories for protein synthesis.

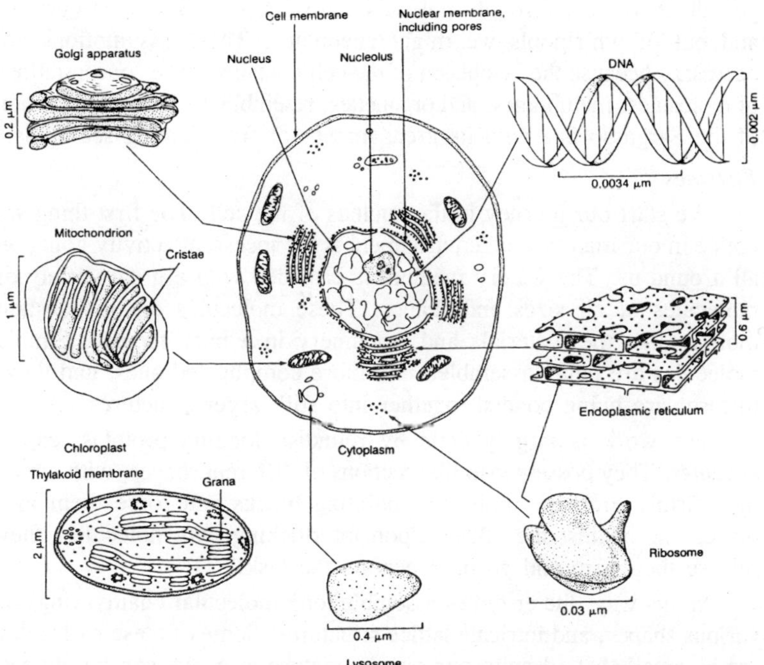

Figure 1.3: The eukaryotic cell. The organelles shown are: Nucleus, Endoplasmic reticulum, Golgi apparatus, Mitochondrion, Chloroplast, Lysosome.

Only eukaryotic cells possess the internal organisation and biochemical sophistication needed to assemble into large and complex multicellular organisms, such as animals and plants. Let us investigate the tasks performed by the organelles and ribosomes of eukaryotic cells. We will discover some of the ways by which the cells function and maintain themselves and will learn how genetic information governs biological processes.

We will be able to compare eukaryotic and prokaryotic cells and gain some appreciation of the enormous differences evolution has created between them.

In other chapter of this book we journeyed to the far reaches of the Cosmos; here we shall take an imaginary journey into the microscopic world inside one of the cells of the human body. The cell is filled with a watery fluid, so we will pretend to explore in a miniature submarine. The sub is about one-tenth of a micrometer in size, which makes it roughly a billion times smaller than a real sub.

Despite its smallness, we assume that our sub is sturdy enough to withstand the dangers ahead and that it has sufficient engine power to permit us to steer through molecular jams as well as against currents and out of whirlpools we might encounter. These assumptions are necessary because the cytoplasm of the cell contains dense concentrations of molecules, membranes, and organelles, resembling a miniature version of the Sargasso Sea with its great masses of free-floating seaweeds.

Enzymes

We start our journey in the nucleus of the cell. The first thing we notice in our microscopic environment is the incessant activity going on all around us. The watery medium is crowded with a great variety of molecules of all sizes and shapes. These molecules are in constant motion, like the materials and machinery in a busy factory. Simple molecules are being assembled into more complicated ones, and they, in turn, are being bonded together into still larger structures.

The work is largely done by roundish looking proteins, called ***enzymes.*** They possess vise-like sections of different shapes, with which they firmly grab the molecular building blocks and force them into place, one against the other. Upon interlocking two molecules, they release their grip and go on repeating the task.

In this way, the enzymes assemble long molecular chains, rings of various shapes, and intricate lattice structures. Some of these molecules are so small that, despite our own miniature size, we can barely see them. Others have sizes and shapes comparable to our own sub. Still others are strands of up to a hundred million times the length of the sub, but only a tiny fraction of its width.

All of these molecules are very stable. When we bump into them, they merely bend and stretch a bit, but they do not break. Only certain enzymes with the correct grip seem to be able to undo the strong bonds holding the molecules together and tear them apart into simpler constituents. Generally this buildup or destruction of molecules is referred to as *a chemical reaction* and the enzymes that facilitate the reactions are *catalysts.*

DNA

The largest molecular structures we find inside the nucleus are DNAs. They are the long, thin molecules mentioned above. They wind back and forth through the nucleus, and it is difficult to understand why they do not become inseparably tangled with each other. Altogether we count 46 of them.

Upon closer inspection, we find that each DNA consists of two parallel strands, connected by single- and double-ringed molecules called *bases*. The two strands wind around each other like those of a rope, forming a double-helical structure with the connecting bases in the middle.

As we examine the structure of a DNA molecule, we notice that the bases, which connect the two strands of the double helix, come in four different versions: *adenine (A), guanine (G), cytosine (C),* and *thymine (T)* (their molecular structures will be described in the following section). The specific sequence of these bases constitutes genetic information, much as the sequences of letters on this page, taken from an alphabet of 26 letters, constitute information. Even though the sequence of bases is continuous, its information content consists of separate units - called *genes* - arranged in a linear sequence one after the other along the DNA.

This is comparable to the subdivision of the continuous sequence of letters in an encyclopedia into separate articles. A DNA molecule may contain thousands of genes. A gene carries one of three different kinds of instructions: (1) assembly instructions for the manufacture of proteins, (2) control instructions for when to make certain proteins, and (3) instructions for making the machinery that assembles proteins (transfer and ribosomal RNA). One of the primary goals of our journey through the cell is to discover how the assembly instructions of the genes are converted into actual proteins.

RNA

Because the genetic instructions are written on molecules of DNA, observing them should give us a start in tracking down the process of protein synthesis. At several places we can see enzymes wedging themselves between the helical strands of DNA and pulling apart short sections corresponding to genes.

This exposes the bases and, hence, the genetic information, and allows other enzymes to copy it. They do that by using molecular raw material present in the nuclear environment and constructing molecules very similar to DNA, except that they consist of only single strands with bases attached to them. These copies are called *RNA,* which stands for *ribonucleic acid.*

The DNA molecules are the "master copy" of genetic information and are kept safely in the nucleus of the cell. The information is transcribed onto RNA molecules for the purpose of translating the information into proteins. There are three different kinds of RNA. One kind, the *messenger RNA* (mRNA), carries information for the assembly of proteins to the protein factories. The other two kinds, the *transfer* and *ribosomal RNA* (tRNA and rRNA), are part of the machinery that assembles proteins.

Let us follow one of the mRNA in the hope that it will take us to a protein factory. For a while, the mRNA meanders through the nucleus. As we steer our sub after it, we pass several DNA and RNA molecules, all sorts of smaller molecules, and large globularly shaped enzymes. Nowhere do we find a protein factory.

Eventually, our mRNA approaches the membrane enclosing the nucleus, which is dotted with tunnels (the technical name is *pore)* leading to the cytoplasm beyond. Through the pores molecular raw materials enter the nucleus and other kinds of molecules leave. Among the molecular products leaving is our mRNA. It seems the protein factories might lie out there in the cytoplasm. We force our way out through one of the pores.

Ribosomes

Outside the nucleus we try to catch up with our mRNA, which is heading toward one of the many globularly shaped bodies scattered through-out the cytoplasm. These bodies are roughly of the size of our sub. They are *ribosomes* and, judging from the activity going on near them, are the factories we have been looking for.

They are surrounded by millions of *amino acids,* the molecular building material of proteins. We count 20 distinct kinds of amino acids, differing from each other in shape and size. Each amino acid is attached to a small RNA molecule, namely a tRNA. Several of the ribosomes have strands of mRNA slowly running through them, much as a computer tape runs through the head of a tape reader, except that the ends of the mRNAs are not neatly wound up on spools.

Simultaneously, the internal machineries of the ribosomes, which consist of enzymes and rRNAs, seize the tRNAs (each with an amino acid attached), line them up, and link the amino acids into growing chains as requested by the instructions. In the process, the tRNAs (now without amino acids) are released.

As amino acid after amino acid gets added and the chains increase in length, they coil into helical structures, which, in turn, fold and twist into very specific configurations. The result is functional protein molecules. It is fascinating to watch the ribosomes at work. They are busy and efficient assembly lines, manufacturing proteins according to instructions delivered by mRNAs.

Usually a single strand of mRNA passes through several ribosomes at once, allowing the simultaneous construction of more than one protein from the same set of instructions. Depending on the number and on the particular sequence of amino acids in the assembled chain, different kinds of proteins result: globularly shaped enzymes for speeding up the chemical reactions in our bodies; proteins that act as chemical sensors or as transporters of certain molecules; and structural proteins for our bones, skin, connective tissues, and the linings of our organs, to name just a few.

Protein synthesis goes on at many places in cells. Some ribosomes exist freely in the cytoplasm. Others are attached to the exterior wall of the nuclear membrane. Still others are associated with vast networks of interconnected channels called the *endoplasmic reticulum*. These channels are used to transport newly manufactured proteins to another set of channels called the *Golgi apparatus* (named after the Italian physician and Nobel laureate Camillo Golgi, who first described this organelle in 1898).

The Golgi apparatus packages the proteins and sends them through the cell membrane to the outside. Thus, the DNAs, RNAs, ribosomes,

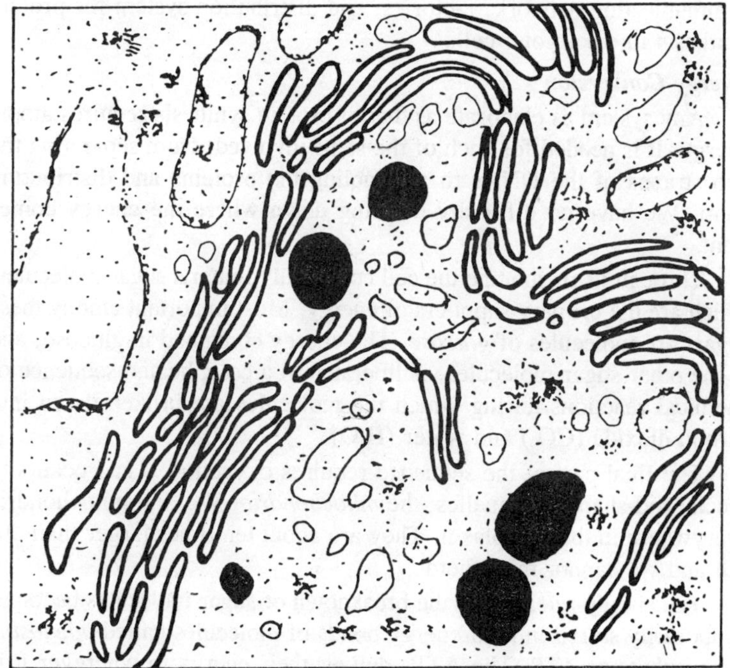

Figure 1.4 : Golgi apparatus from the green alga Chlamydomonas (× 78,000). Note the characteristic stacks of parallel, folded membranes of the apparatus.

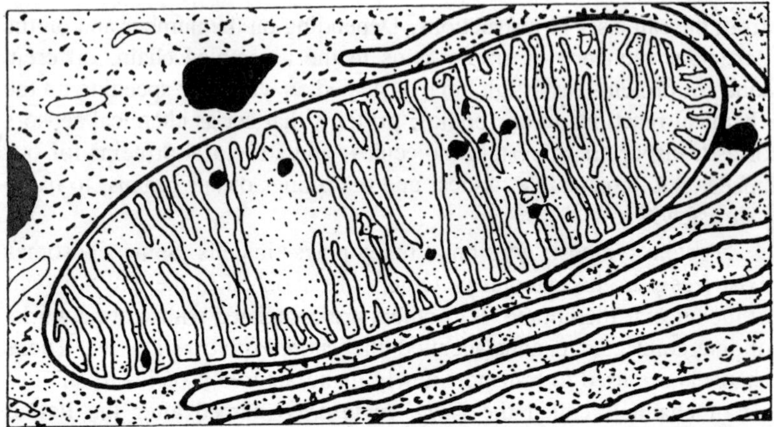

Figure 1.5 : Mitochondrion from the cell of a bat (× 50,000). The mitochondrion's double outer membrane and numerous folded inner membranes (the cristae) are clearly visible. Also visible, outside the mitochondrion, is a section of the cell's endoplasmic reticulum, with ribosomes attached to its membrane.

endoplasmic reticulum, and Golgi apparatus constitute the management, communication network, factories, and distribution system for protein synthesis in eukaryotic cells.

Energy Conversion

Any system as elaborate as that of protein synthesis requires ample energy. It is needed for each of the steps involved, from separating the two strands of the DNAs to assembling the proteins and distributing them. We have no difficulty tracking down where the energy comes from.

Through its membrane the cell continually imports sugar molecules, which are the primary suppliers of energy. Most important among these sugars are molecules of *glucose*. The energy contained in glucose, and in the other sugar molecules, is liberated in a complicated sequence of chemical reactions during which the molecules are broken down into carbon dioxide (CO_2) and water (H_2O).

The final part of the sequence requires oxygen (O_2) and occurs in special rod-shaped organelles, the *mitochondria*. We find mitochondria everywhere in the cytoplasm. They are about ten times larger than our sub and we cannot miss them.

The energy released by the breakdown of sugar molecules becomes temporarily stored in high-energy bonds of molecules called *adenosine triphosphate* or *ATP*. The ATPs deliver their energy to wherever it is needed by the cell, such as in the various steps of protein synthesis or the transport of materials within cells and across cell membranes. (Contraction of muscle cells and conduction of nerve impulses are still other examples of processes utilizing energy stored in ATP) The full sequence of chemical reactions by which cells convert the energy of sugars into high-energy bonds of ATP with the participation of O_2 is called *aerobic respiration*.

If we were exploring a cell from the green leaves of plants or from algae instead of one from our own body, we would find chloroplasts in addition to mitochondria and the other organelles. *Chloroplasts* ("green bodies") are disk-shaped bodies that harness energy by *photosynthesis*. They trap the energy of sunlight with the aid of molecules of chlorophyll and other similar pigments.

Like the mitochondria, they temporarily store energy in high-energy bonds of ATP However, instead of requiring $_{02,}$ they release it. They then use the ATPs to run their metabolic machinery. In particular, they use them to convert CO2 and H_2O into energy-rich sugars and other carbohydrate molecules. Thus, as described more fully in other chapter of this book, photosynthesis and respiration are parts of the same cycle.

In photosynthesis, energy from the Sun is used to elevate the carbon, oxygen, and hydrogen atoms of CO_2 and H_2O into energy-rich sugar molecules. Molecular oxygen is expelled in the process. In aerobic respiration, the sugars are broken down, with the participation of O_2, into the low-energy molecules CO_2 and H_2O, and the energy is put to use.

We have now completed our tour through the eukaryotic cell and are ready to return to our normal macroscopic environment. We do this

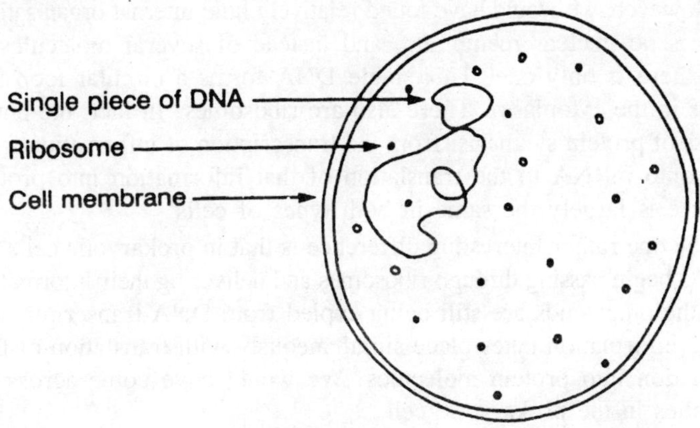

Figure 1.6 : The prokaryotic cell.

by disguising ourselves as a protein molecule manufactured by the cell and leaving via the Golgi apparatus. We must be careful, however. In the vicinity of the Golgi apparatus we are likely to encounter *lysosomes,* which are roundish saclike organelles filled with very potent enzymes.

The lysosomes attack foreign objects, such as bacteria, viruses, or defective proteins, and engulf them. Within minutes, the enzymes break the objects down into simple molecular units, which are then recycled by the cell.

Because we do not wish to test how well our sub would stand up to such an attack, we allow the Golgi apparatus to wrap us into a membrane coat, as it does with proteins ready for export. This makes us indistinguishable from proteins manufactured by the cell and we safely escape through the cell membrane to the outside.

Prokaryotes

In contrast to the eukaryotic cells, most prokaryotic cells have sizes in the range from 1 to 5 μm. But there are exceptions. The smallest bacteria, the mycoplasmas, measure only about 0.3 μm across, while a few of the blue-green bacteria (also known as cyanobacteria, and formerly called blue-green algae, though they are not algae) may grow as large as small eukaryotic cells. Some prokaryotes aggregate into single-stranded and branched filaments or into globular colonies.

However, none show evidence of cell specialisation, nor do they generally depend on each other for survival. Had we toured a prokaryotic cell, we would have found that, like the eukaryotic cell, it is filled with cytoplasm and bounded by a cellular membrane, through which it exchanges materials with the surroundings.

However, we would have found relatively little internal organisation. There is no nuclear membrane, and instead of several molecules of DNA there is only one. This single DNA forms a circular loop and resides in the cytoplasm. There also are ribosomes. In fact, the entire process of protein synthesis, from the transcription of information from DNA onto mRNA to the translation of that information into protein structure is largely the same in both types of cells.

The one rather interesting difference is that in prokaryotic cells the mRNAs begin passing through ribosomes and delivering their information while the other ends are still being copied from DNA-transcription of genetic information takes place simultaneously with translation of that information into protein molecules. We would have come across no organelles in the prokaryotic cell.

In their absence, ATPs are produced on the cell's enclosing

membrane and molecular materials are transported directly through the cytoplasm without the benefit of special channels.

These descriptions of eukaryotic and prokaryotic cells point to the profound differences existing between them. Eukaryotic cells are unquestionably more complex than prokaryotic cells. Nevertheless, important similarities also exist between them, which suggests that eukaryotic cells evolved from prokaryotic ancestors.

That this is in fact the case becomes apparent when the biochemical processes of the two types of cells are compared. The structures of DNA and RNA are nearly identical in both cells and the processes of cell division in eukaryotic cells seem to be derived from those in prokaryotes. Both cells use similar steps in the synthesis of proteins. Furthermore, parts of the metabolic pathways of the eukaryotic cells are closely related to those of prokaryotes. The next two chapters will investigate these rather interesting relationships.

ORGANIC MOLECULES : THE FUNDAMENTAL BUILDING BLOCKS OF LIFE

This discussion of cell structure and function indicates that life's fundamental building blocks are organic molecules. The other constituents —water, ions, and trace elements-are mere support units. Organic molecules are so special for life because they possess a combination of rather unique properties:

1. They come in a number of small basic units that can be linked together into many different kinds of large and elaborate structures, such as chains, rings, lattices, long fibers, and globules.
2. Because of their diverse structures, organic molecules are able to participate in complex, self-regulating chemical reactions. These reactions are the source of all biological processes.
3. The chemical bonds holding organic molecules together are quite strong and not easily broken. That is why these molecules form and persist in such dissimilar environments as interstellar clouds, meteorites, the atmospheres of the Jovian planets, and organisms.
4. The primary constituents of organic molecules—carbon, hydrogen, oxygen, nitrogen, phosphorus, and sulfur—are among the most abundant elements on the surface of our planet.

Organic molecules are made by joining carbon with carbon, hydrogen, oxygen, nitrogen, and other atoms by chemical bonds. Carbon is always connected by four bonds to other atoms, nitrogen by three, oxygen by two, and hydrogen by one. Examples of these bonds are found in methane (CH_4), ammonia (NH_3), and water (H_2O), which were described in the discussion of interstellar clouds.

The bonds between carbon and other atoms need not necessarily be single bonds, as in the case of methane. They may be double or triple bonds. For instance, in formaldehyde carbon forms a double bond with oxygen,

$$\begin{array}{l} H \diagdown \\ \quad\;\; C=O \\ H \diagup \end{array}$$

and in acetylene (a gas that burns brightly in the presence of oxygen and is used in welding and soldering) two carbon atoms are joined by a triple bond,

$$H—C \equiv C—H.$$

This characteristic of carbon—to form strong single, double, and triple bonds—gives organic molecules their great versatility and, together with the four properties discussed above, makes them the natural and, possibly, only workable choice for the construction of organisms.

Although organic molecules assume a virtually unlimited variety of structures, many of these structures exhibit certain similarities. This allows chemists to group organic molecules into distinct classes and greatly simplifies the way we think about them.

The larger, so-called polymeric (meaning "many parts") organic molecules that are of biological significance may be conveniently placed into four major classes: the carbohydrates, lipids, proteins, and nucleic acids. The *carbohydrates* comprise the starches we eat and from which we derive much of our energy, the cellulose that gives wood its rigidity, and the glycogen by which we store energy in our bodies.

The *lipids* include the fats of our bodies, the cholesterol in our blood, and many of the hormones, such as testosterone and estrogen (male and female sex hormones). The fats are primarily used for storage of energy and for the structural envelopes of cells and their constituent organelles.

The *proteins* serve several functions. They may be structural proteins, giving our bones their strength, lining our organs, and holding our bodies together through their presence in skin, tendons, and connective tissues. Structural proteins are also associated with certain fats (see above) to form the membranes of cells and their organelles. Other proteins are chemical sensors and transporters of certain molecules. Still other proteins act as enzymes and are responsible for the smooth functioning of all chemical reactions going on in the cells, including reproduction, growth and self-maintenance. The *nucleic acids* are the DNAs and RNAs, which are carriers of genetic information.

Carbohydrates

The *carbohydrates* consist of carbon, hydrogen, and oxygen and are constructed by combining carbon with the atomic constituents of water. The sugars are the simplest carbohydrates, and the most important sugar molecules are *glucose, ribose,* and *deoxyribose.* The glucose molecule is one of the suppliers of energy in our cells.

You may recognize ribose and deoxyribose, for the names ribo- and deoxyribonucleic acids indicate that these two sugars are present in RNA and DNA. Another sugar molecule is *fructose.* Bonded to glucose, it forms sucrose, ordinary table sugar. Glucose, fructose, ribose, and deoxyribose are ring-shaped molecules.

The chemical formula of glucose is $C_6H_{12}O_6$, meaning that it is constructed of six carbon, twelve hydrogen, and six oxygen atoms. It is manufactured during photosynthesis from CO_2 and H_2O, with the simultaneous release of O_2:

$$6\ CO_2 + 6\ H_2O \rightarrow C_6H_{12}O_6 + 6\ O_2.$$

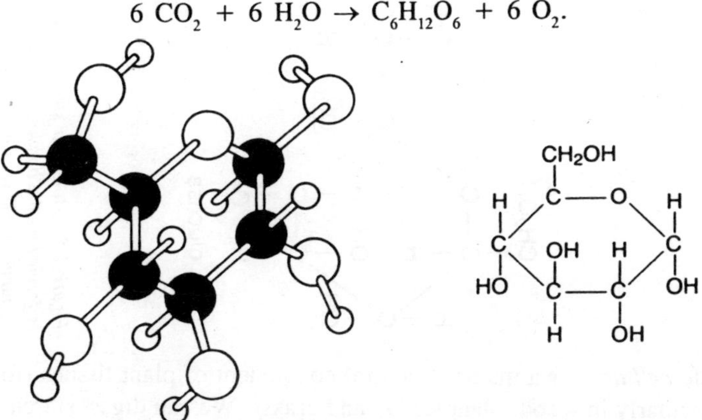

(a) Three-dimensional model (b) Two-dimensional model

Figure 1.7 : Two ways of representing the glucose molecule. A third way is by the chemical formula $C_6H_{12}O_6$.

Glucose molecules are then assembled into long and often branched chains. Depending on the particular way the glucose molecules are bonded to each other, the result is either starch or cellulose. *Starch* is a major constituent of vegetables, fruit, and some roots (carrots, potatoes);

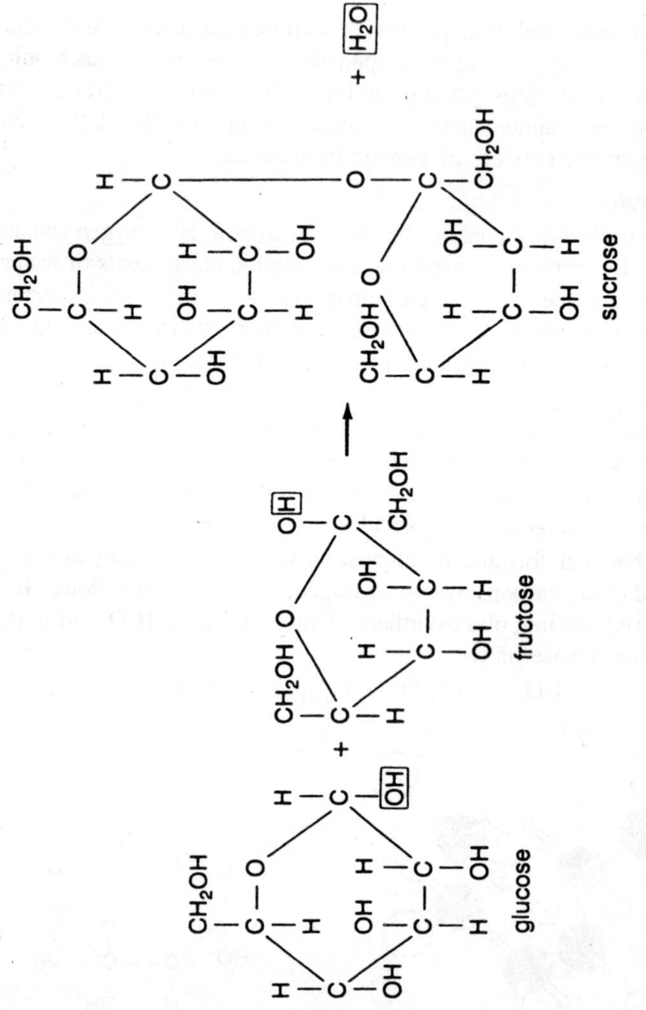

Figure 1.8 : Molecular structure of sucrose. Ordinary table sugar consists of crystals of sucrose molecules. Sucrose is formed by the bonding together of glucose and fructose, with the simultaneous removal of a water molecule (see rectangles). Glucose and fructose are ring-shaped carbohydrate molecules with the same chemical formula, $C_6H_{12}O_6$. *However, the ring of glucose is hexagonal, that of fructose is pentagonal.*

while *cellulose is* a major structural component of plant tissues (found particularly in wood, plant stalks, and grass). We can digest starch, for our bodies produce the enzymes required to break the bonds between its glucose molecules. We cannot digest cellulose because we do not have the necessary enzymes, but insects, cows, and many other animals can feed on it. A part of their digestive tract is inhabited by microorganisms (bacteria and protists) that produce enzymes capable of breaking down cellulose.

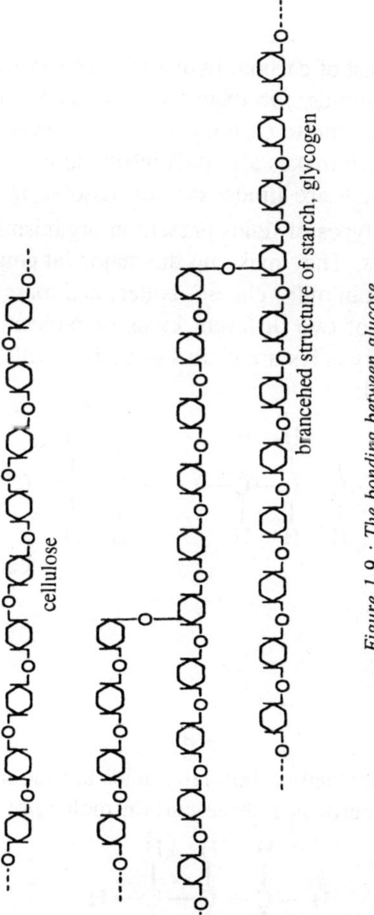

Figure 1.9 : The bonding between glucose molecules in cellulose, starch, and glycogen.

Upon digestion of starch, individual glucose molecules enter our blood stream and are carried to the cells to deliver their energy. That energy is converted into high-energy bonds of ATP, as discussed in the previous section. Excess glucose enters the liver and is stored for future use in long molecular chains identical to starch, except that they are more highly branched.

These molecular chains are called *glycogen*. Whenever the glucose concentration in our blood—also known as blood sugar—runs low, such as with strenuous exercise, we feel tired. Glycogen is then broken down in the liver and glucose molecules are released into the bloodstream for transport to the active cells.

Lipids

The *lipids* consist of carbon, hydrogen, and oxygen (though some of them, the phospholipids for example, also contain phosphorus and nitrogen). In comparison to carbohydrates, the oxygen content of lipids is much lower, which makes all lipids insoluble in water. For instance, grease and oil (which are lipids) do not dissolve in water.

Of the several types of lipids present in organisms, here we discuss only the *neutral fats*. They make up the major fat content of our bodies and are also present in milk, cheese, butter, and margarine. The neutral fats are composed of two different kinds of molecular units, *glycerol* and *fatty acids*. Fatty acids are chains of carbon with a carboxyl group* attached at one end:

$$\begin{array}{ccccccccccccccccc}
 & & H & & H & & H & & H & & H & & H & & H & & \\
 & & | & & | & & | & & | & & | & & | & & | & & {}^{\diagup\!\!\diagup} O \\
H & - & C & - & C & - & C & - & C & - & C & -\cdots- & C & - & C & - & C \\
 & & | & & | & & | & & | & & | & & | & & | & & {}_{\diagdown} O - H \\
 & & H & & H & & H & & H & & H & & H & & H & &
\end{array}$$

Carboxyl group

*The *carboxyl group*,

$$\begin{array}{ll}
 & {}^{\diagup\!\!\diagup} O \\
-C & \\
 & {}_{\diagdown} O - H
\end{array}$$

Sixteen- and eighteen-carbon fatty acids are the most common ones in our bodies. Glycerol is a three-carbon molecule:

$$\begin{array}{ccccccccc}
 & & H & & H & & H & & \\
 & & | & & | & & | & & \\
H & - & C & - & C & - & C & - & H \\
 & & | & & | & & | & & \\
 & & O & & O & & O & & \\
 & & | & & | & & | & & \\
 & & H & & H & & H & &
\end{array}$$

Neutral fats are also known as *triglycerides* because they consist of glycerol linked with three fatty acids. During medical checkups, it is now common to have one's blood examined for triglyceride concentration.

The atomic constituents of the water molecules liberated during the formation of neutral fats come from both the glycerol and the fatty acids, as shown by the shaded rectangles in the above chemical reaction. Such *dehydration* reactions are the usual way of assembling large organic molecules (polymers) from basic units (monomers). During the breakdown of polymers, the process is reversed and water must be supplied. The reverse reactions are called *hydrolysis,* meaning that they require the splitting of water into H and OH. *(Hydrolysis is* derived from the Greek *hydro-* and *-lysis,* meaning "water" and "loosening" or "splitting." The root *hydro- is* also present in the word *carbohydrate.*)

To make neutral fat molecules, three fatty acids are linked to a glycerol, with the simultaneous removal of three water molecules:

```
  H            O H      H          H     O H      H
  |            ‖ |      |          |     ‖ |      |
H—C—O—H + H—O—C—C— ··· —C—H      H—C—O—C—C— ··· —C—H
  |              |      |          |       |      |
  |              H      H          |       H      H
  |            O H      H          |     O H      H
  |            ‖ |      |          |     ‖ |      |
H—C—O—H + H—O—C—C— ··· —C—H  →   H—C—O—C—C— ··· —C—H + 3 H2O
  |              |      |          |       |      |
  |              H      H          |       H      H
  |            O H      H          |     O H      H
  |            ‖ |      |          |     ‖ |      |
H—C—O—H + H—O—C—C— ··· —C—H      H—C—O—C—C— ··· —C—H
  |              |      |          |       |      |
  H              H      H          H       H      H
```

Glycerol ↑ 3 Fatty acids Neutral fat (triglyceride) ↑

−H and H—O− are removed and combined into molecules of water

3 Molecules of water

In some neutral fats, all the carbon atoms of the fatty acids are joined by single bonds, as in the examples above. These fatty acids contain the maximum possible number of hydrogen atoms and, therefore, are called *saturated.*

In other fats, some of the carbon atoms of the fatty acids possess only one H and are linked to a neighboring carbon atom by a double bond:

```
   H   H   H   H           H   H
   |   |   |   |           |   |    O
   |   |   |   |           |   |   //
H— C — C — C — C — C ═ C — C — C — C
   |   |   |   |   |   |   |   |   \
   |   |   |   |   |   |   |   |    O—H
   H   H   H   H   H   H   H   H
```

Such fats are said to be *unsaturated* or, if there are several such double bonds, *polyunsaturated.* Animal fats are mostly saturated, while those of vegetables, seeds, and nuts are usually polyunsaturated. There seems to exist a strong correlation between consumption of saturated fats with fatty deposits in the blood vessels and heart disease. Hence, nutritionists recommend that we eat only moderate amounts of meat and diary products.

Proteins

Proteins contain nitrogen and, in some instances, sulfur in addition to carbon, hydrogen, and oxygen. Of all the organic molecules found in organisms, the proteins have the most varied range of applications. Not only are they the major *structural materials* of organisms, they also act as *enzymes,* which catalyze all biochemical reactions from those of bacteria to those of man.

Still other proteins serve as *hormones* or as *carrier molecules* in animals. An example of a protein hormone is insulin. It is produced in the pancreas (a gland) and assists in the transport of glucose across cell membranes, thereby speeding up the rate of energy use by cells. An example of a protein carrier molecule is hemoglobin.

Proteins are polymers, consisting of large numbers of amino acids (the monomers) bonded together in long linear chains. Of the hundreds of conceivable amino acids, only twenty occur in the protein of organisms. The three simplest amino acids are glycine, alanine, and serine:

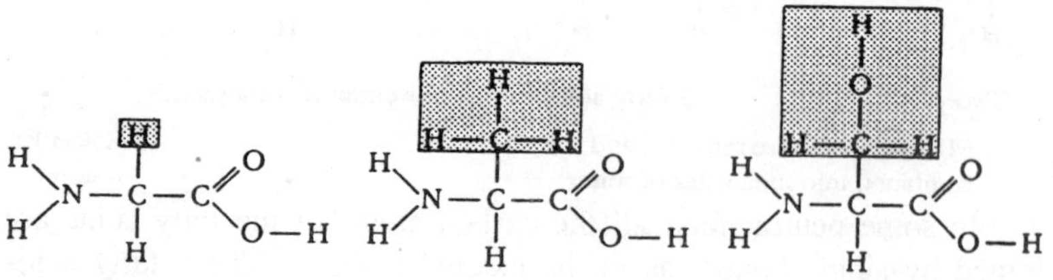

Glycine Alanine Serine

Like all amino acids, these three possess *a carboxyl group*, an *amino group* and a molecular *side chain* that is unique for each amino acid and distinguishes it from the rest. In glycine the side chain (marked by shaded rectangles in the above formulae) is a single hydrogen atom (H), in alanine it is CH_3, and in serine it is CH_2OH. If we label the side chains by the letter R, we obtain a generalised formula for amino acids:

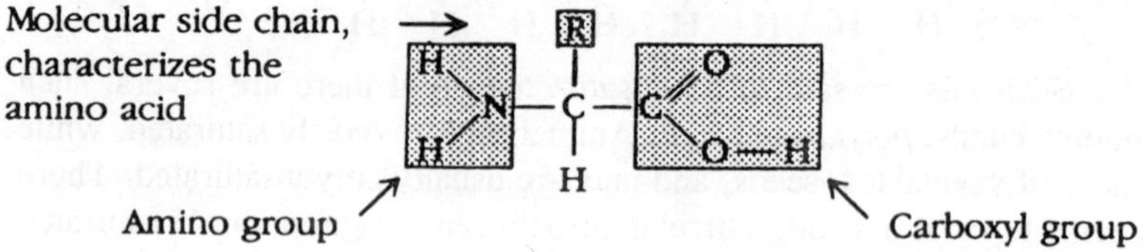

Molecular side chain, characterizes the amino acid →

Amino group ↗ ↖ Carboxyl group

Figure elsewhere in this chapter lists all twenty amino acids occurring in terrestrial life. In addition to carbon, hydrogen, oxygen, and nitrogen, two of the amino acids (cysteine and methionine) contain sulfur. In most amino acids, the side chains are linear or branched, but

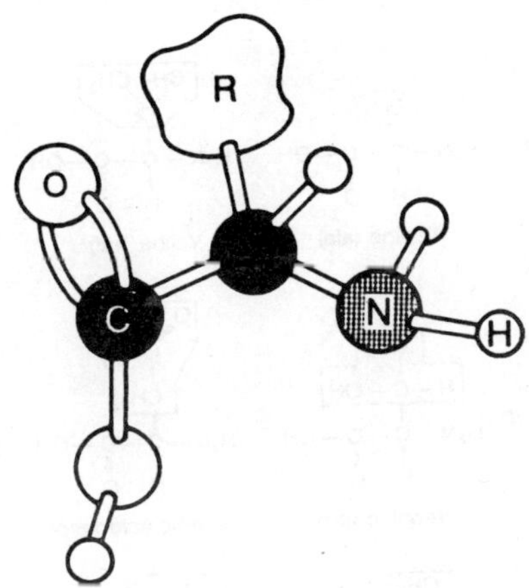

Figure 1.10 : Threedimensional model of an amino acid. R designates the molecular side chain.

in the last five amino acids shown in figure elsewhere in this chapter they are ring-shaped. Unlike other life forms, animals have lost the ability to synthesize all of the twenty amino acids. For example, adult humans do not synthesize valine, leucine, isoleucine, threonine, lysine, methionine, phenylalanine, tryptophan, and histidine. We must obtain these nine "essential" amino acids from our diet.

The assembly of amino acids into linear chains occurs by dehydration reactions, like those involved in the making of cellulose, starch, glycogen and neutral fats. The OH from the carboxyl group of one amino acid and an H from the amino group of a second amino acid are removed (as H_2O) and the C and N thus exposed are bonded together:

$$\underset{\text{First amino acid}}{H_2N{-}CH(R_1){-}C(=O){-}OH} + \underset{\text{Second amino acid}}{H{-}NH{-}CH(R_2){-}C(=O){-}OH} \rightarrow \underset{\text{Peptide}}{H_2N{-}CH(R_1){-}\underbrace{C(=O){-}NH}_{\text{Peptide bond}}{-}CH(R_2){-}C(=O){-}OH} + H_2O$$

A third, fourth, and still other amino acids are then added, until chains containing hundreds and even thousands of amino acids are obtained. Such chains are called *peptides* (they are not yet functional proteins), and the bonds linking the amino acids are called *peptide bonds*.

In cells the assembly of peptides is far more complicated than indicated by the above description. Not only has a peptide bond to be formed, but it has to be formed between two particular amino acids. The instructions for the sequence in which particular amino acids are

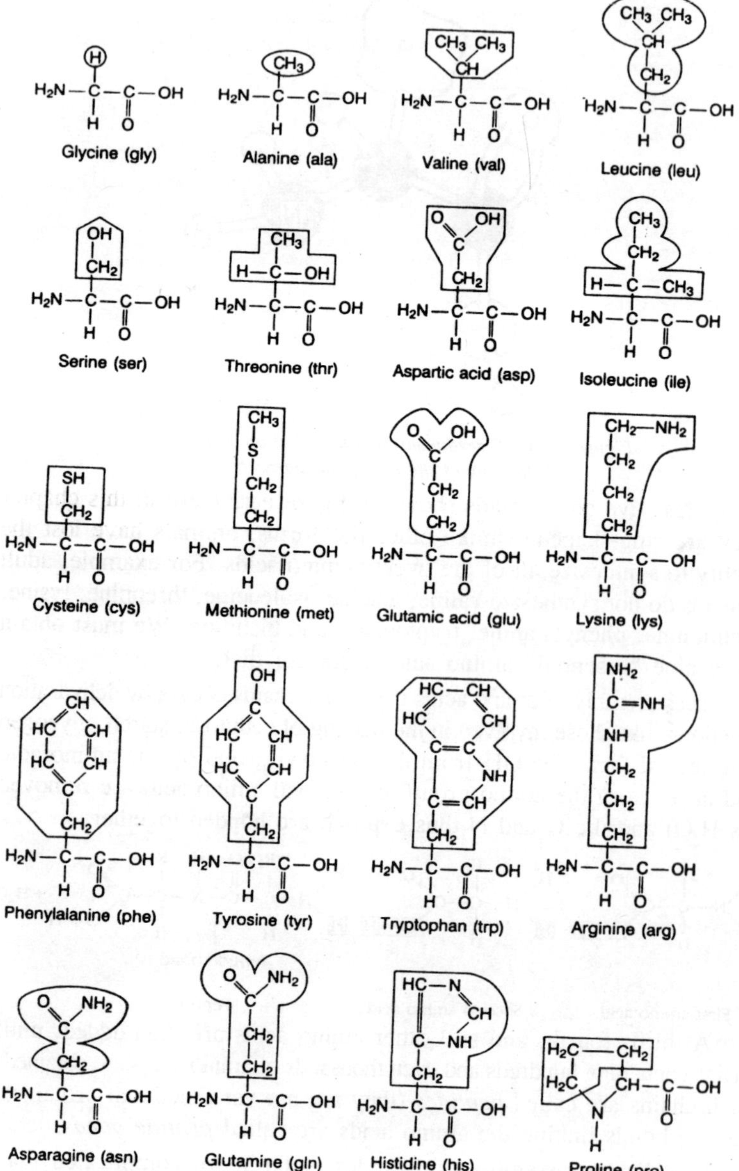

Figure 1.11 : The twenty amino acids present in protein.

to be selected from the twenty existing in organisms and strung together into a peptide chain comes ultimately from the cell's DNA—the storage molecule of genetic information. The translation of these instructions from DNA into peptides involves RNAs and ribosomes and is a process that is central to what makes organisms alive.

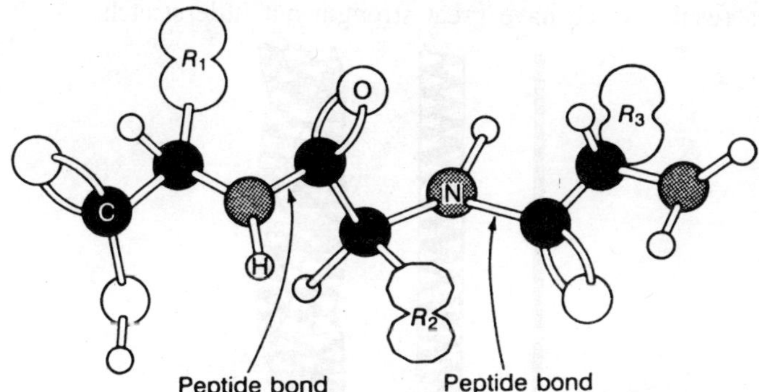

Figure 1.12 : A short peptide chain, consisting of three amino acids. The amino acids are linked by peptide bonds.

Linear chains of peptides represent the *primary* structure in the hierarchy of the structures that constitute a protein molecule. As a peptide chain grows, it forms a coiled strand that is the *secondary* structure in the hierarchy. Depending on the precise sequence of amino acids, the coiled peptide strand folds back and forth on itself, forms knot- or pretzellike configurations, or twists into a spiral.

Such bending, folding, or spiraling represents the *tertiary* structure. Generally, tertiary peptide structures do not occur alone, but are intertwined with other tertiary peptide structures. The result is the *quaternary* structure in the hierarchy and constitutes a functional protein molecule.

The twisting and folding of peptide chains into functional proteins produce molecules that are either long and fibrous or compact and globular. Examples of these two types of proteins are collagen (fibrous) and hemoglobin (globular). *Collagen* is an important structural component of our bodies.

Its synthesis begins with the assembly of amino acids, most of which are glycine and proline, into long peptide chains. Upon coiling into secondary structures, three of the peptides wind around each other into *tropocollagen* molecules.

The tropocollagen molecules are then joined in staggered fashion to form a collagen fibril (tertiary and quaternary structures), much like single strands of wool fibers are spun into yarn. From these fibrils still larger structures are assembled, analogous to the weaving or knitting of yarn into finished woolen products. If the collagen fibrils are combined into long parallel bundles, structures, like tendons and other connective tissues result, which have great strength but little stretch.

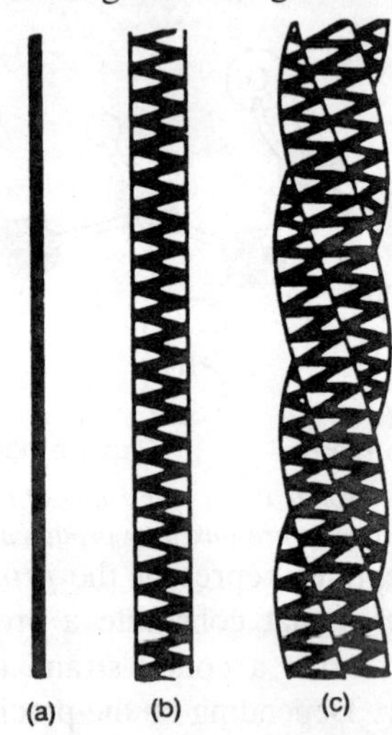

Figure 1.13: The structure of protein: (a) Diagram of a linear peptide chain - primary structure; (b) a coiled peptide chain -secondary structure; (c) three coiled peptide chains twisting around each other to form a fibrous protein, as, for example, a fiber of collagen. The twisted form of a single coiled peptide chain represents the tertiary structure. The combination of the three peptide chains twisting around each other represents the quaternary structure and constitutes a protein molecule.

If they are meshed into flat, interlacing networks, human skin or animal hide is formed, which are highly flexible. If the fibrils are interwoven with calcium-containing inorganic molecules (for example, calcium phosphate and calcium carbonate), bone is made, which is strong and hard without being brittle.

Hemoglobin is contained in red blood cells and is the carrier molecule of O_2 in blood. It consists of four coiled peptide chains, which are tightly folded together forming the *globin* part of hemoglobin. In its midst, each of the peptide chains holds a relatively small, disk-shaped *heme* molecule (which is not a protein) that contains an atom of iron. The iron gives blood its deep red colour.

As blood flushes through the capillaries of the membranes lining the lungs, hemoglobin picks up O_2. The O_2 (bound to hemoglobin) is then carried by the bloodstream to the body tissues, where it is released. (The word *heme* is derived from the Greek and means "blood." *Globin*

Figure 1.14 : Collagen fibrils, carefully pulled away from human skin (× 42,000). The distinct dark bands visible on the fibrils are due to a partial overlapping of adjacent tropocollagen molecules and are about 0.07 μm apart.

comes from the Latin *globus,* meaning "ball.") The folding of hemoglobin is typical of all globular proteins. The nature of the folding and, hence, the uses to which the proteins are put -enzymes, hormones, carriers - depends on the number and precise sequence of amino acids in the peptide chains. As discussed next, this number and sequence of amino acids is determined by the genetic information carried in coded form by DNA.

Nucleic Acids

The *nucleic acids* are made of carbon, hydrogen, oxygen, nitrogen, and phosphorus. Examples of nucleic acids are DNA and RNA. The major functions of DNA and RNA are to store genetic information in coded form, to participate in the conversion of this information into protein structure, and to pass on the information from generation to generation.

DNA generally consists of two molecular strands that spiral around each other like the two strands of a rope. That is why James D. Watson

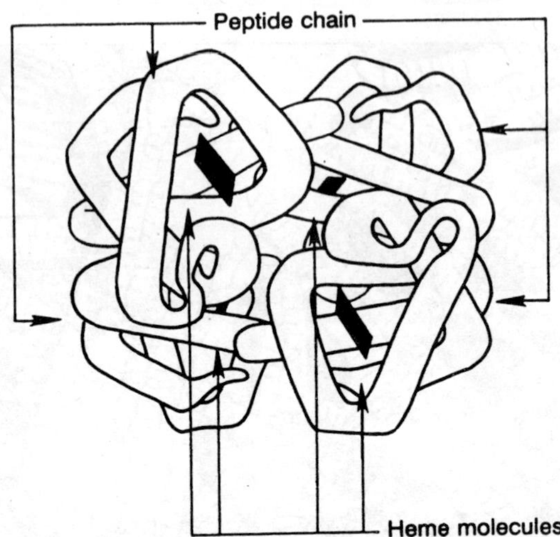

Figure 1.15 : Hemoglobin. The folding, twisting, and intertwining of the four coiled peptide chains of hemoglobin illustrate the quaternary structure of globular protein molecules.

and Francis H. C. Crick, the discoverers of its molecular structure, referred to it as the "double helix". *RNA* usually consists of a single molecular strand, whose structure differs only in minor ways from that of an individual strand of DNA.

Each of the strands of DNA or RNA is a polymer, whose monomers are called *nucleotides.* The nucleotides consist of three basic molecular units -a phosphate, a sugar, and a base.

The *phosphate* is an inorganic molecule. The sugar molecule is *deoxyribose* in the case of DNA and *ribose* in the case of RNA (see above discussion of carbohydrates). The bases of DNA are *adenine (A), guanine (G), cytosine (C),* and *thymine (T).* The bases of RNA are the same as those of DNA except that thymine is replaced by a similar base called *uracil (U).* The bases A and G are double-ringed and called *purines.* The bases C and T (U) are single-ringed and called *pyrimidines.*

To make a strand of DNA or RNA, the nucleotides are strung together such that a phosphate is bonded to a sugar, the sugar is bonded to another phosphate, and so on, with the bases sticking off to the side.

The sugar-phosphate sequence is often called the "backbone" of the strand. The sequence of bases, carried by the backbone, expresses genetic information.

In the case of DNA, the two molecular strands run parallel to each other. The bases of the two strands point toward each other and are weakly bonded together by *hydrogen bonds*, like the interlocking teeth

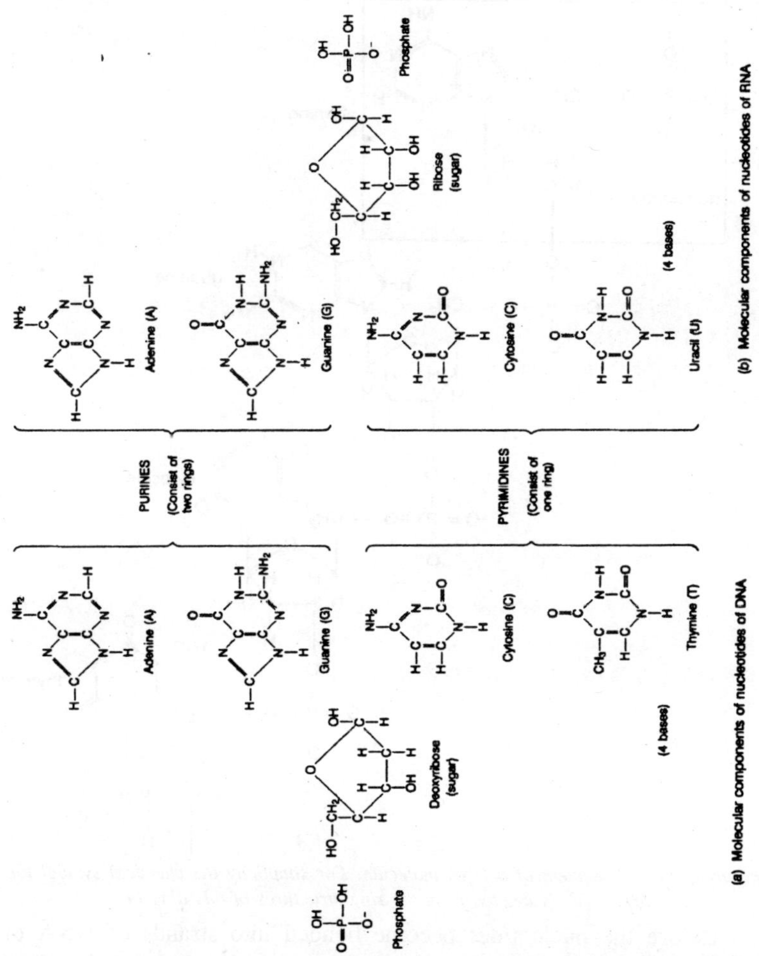

Figure 1.16 : Molecular components of nucleotides.

of a closed zipper. Simultaneously, the two strands wind around each other in a double helix.

The bases of one strand of DNA are always bonded to the corresponding bases of the other strand according to the rule: A-T and G-C (or T—A and C—G). This ensures that a single-ringed base is always bonded with a doubleringed base, making the separation between the two sugar-phosphate chains the same all along the DNA molecule and providing for a neat, parallel structure. More important, this arrangement is the key to accurate replication and usage of genetic information, discussed further in other chapter of this book.

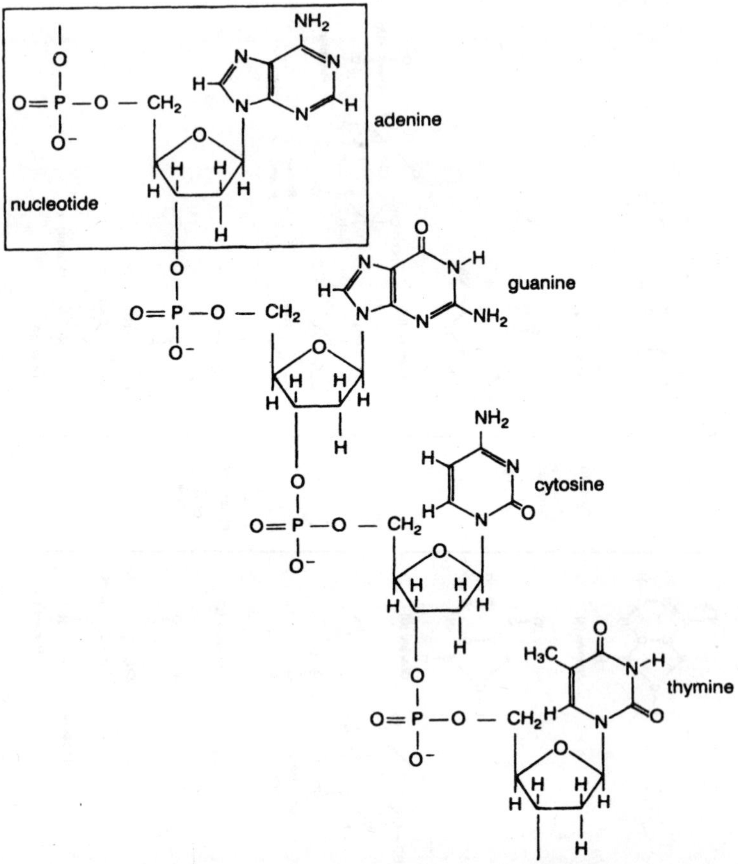

Figure 1.17 : A segment of a DNA molecule. For simplicity the chemical symbol for carbon (C) is deleted from the ring structures of this diagram.

Before the nucleotides become bonded into strands of DNA or RNA, they exist in a form just slightly more complex than that described above. Instead of a single phosphate, they possess three phosphate molecules:

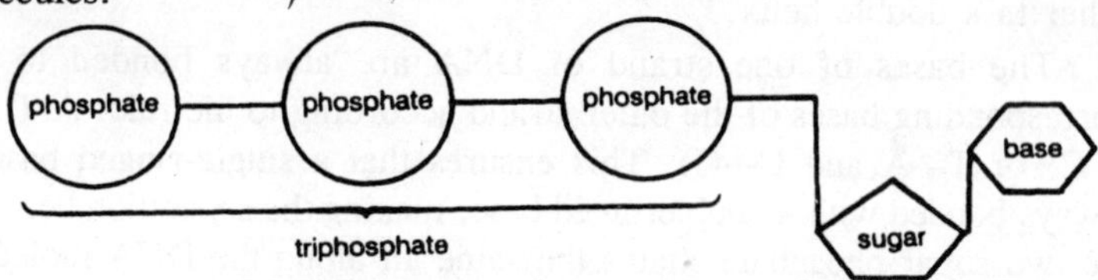

This arrangement of three linked phosphates is known as a *triphosphate,* and nucleotides possessing a triphosphate are said to be

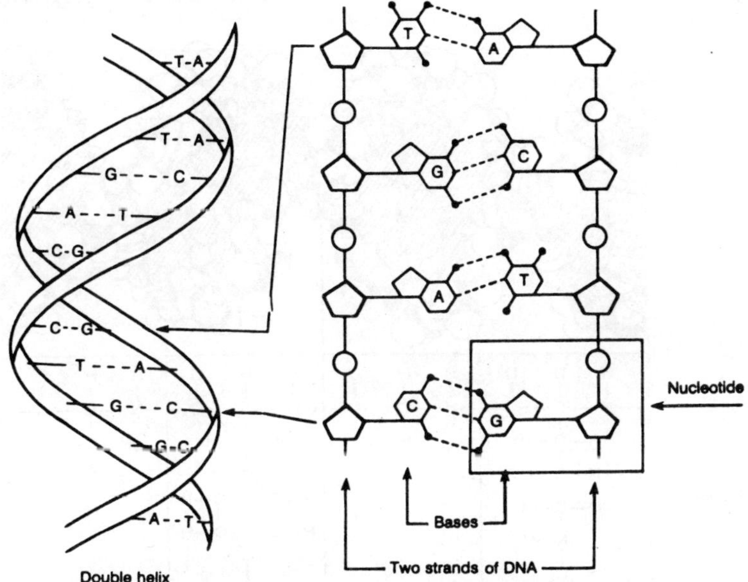

Figure 1.18 : The molecular structure of DNA. The dashed lines represent the hydrogen bonds between bases A and T, G and C.

activated. For reasons connected with the bond structure linking the phosphates, a triphosphate is at an unusually high energy level.

The energy is released when the bonds between the phosphates are broken. This is what happens when the nucleotides are assembled into strands of DNA and RNA. Two of the phosphates are broken off and the energy thus released is used to form the new bond between the single remaining phosphate and the sugar of the next nucleotide.

The principle for energy storage and energy use just discussed has been adopted by life for its most common energy-carrying *molecule-adenosine triphosphate* or ATP. ATP is nothing but a nucleotide with a triphosphate chain. Its sugar is ribose and the base is adenine:

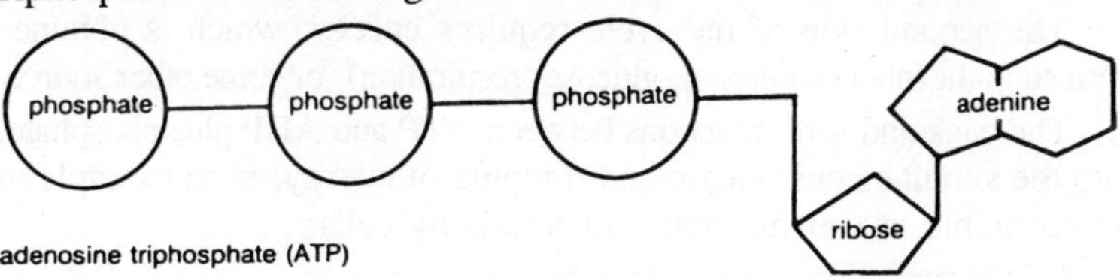

By breaking off one of the phosphates (by hydrolysis), ATP is changed into *adenosine diphosphate* or *ADP* and energy is released for use by the cells. The change from ATP to ADP with the simultaneous

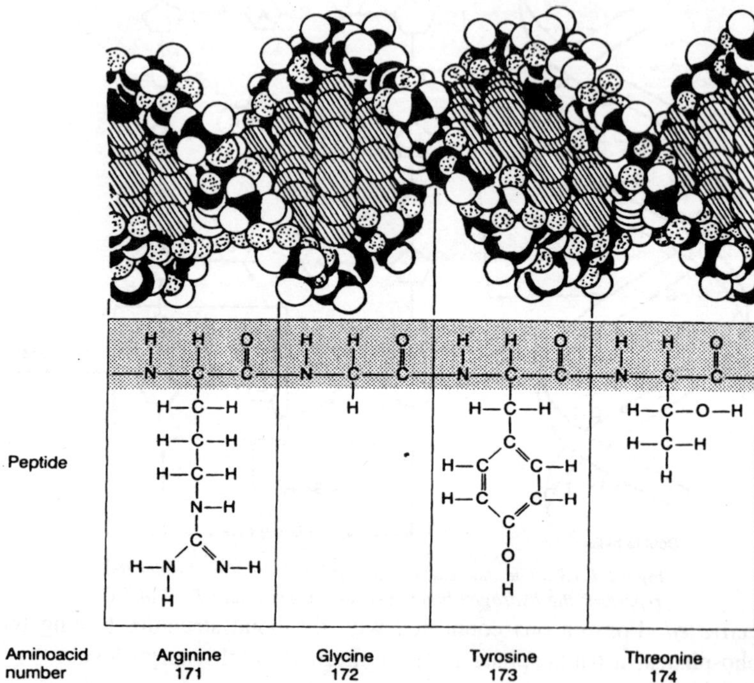

Figure 1.19 : Scale drawing of a section of a DNA molecule (top) and the corresponding peptide chain (bottom). The linear sequence of bases carried by DNA constitutes the instructions for the assembly of a linear sequence of amino acids, which constitutes a peptide chain.

release of energy is one step in a dual cycle. In the second step a phosphate is added to ADP, converting it back to ATP (a dehydration reaction):

Release of energy: ATP + H_2O → ADP + phosphate + energy

Storage of energy: ADP + phosphate + energy → ATP + H_2O

The second step of the cycle requires energy, which is obtained from sunlight (photosynthesis), glucose (respiration), or some other source.

The back and forth reactions between ATP and ADP plus phosphate, with the simultaneous release and trapping of energy, is an example of the economic use of molecular materials by cells.

It is a recycling process in which the same molecules (ADP and phosphate) are used over and over again, while energy flows through the system and drives it. Similar recycling processes are found in many of the chemical pathways of living organisms.

On a more global scale, the flow of carbon, hydrogen, oxygen, nitrogen, and the other biologically important elements through the Earth's ecosystem also constitutes a recycling process. These elements become incorporated into biological organisms, only to be returned eventually to the ground, the water, or the air. As in the example of ADP and phosphate, this larger cycle requires the continuous input of energy, which ultimately comes from the Sun.

THE GENETIC CODE

In the previous section we discussed about the molecular structures of DNA and RNA. But how does the sequence of bases in these molecules express genetic information? Focus on one of the strands of DNA and note that it carries its bases in a linear sequence, much as one side of a zipper carries its teeth in a linear sequence.

There is one important difference, however, between the sequence of teeth of a zipper and the sequence of bases in a strand of DNA. The teeth of a zipper are all alike and hence contain no information. In contrast, there are four DNA bases—adenine, guanine, cytosine, and thymine, or A, G, C, and T (A, G, C, and U [uracil] in the case of RNA). Depending on how the four bases are arranged in sequence, a great many different kinds of messages can be written, just as many different messages can be written with the 26 letters of the alphabet.

Recall that DNA carries only three kinds of instructions: *assembly* instructions for how to make proteins (which involves mRNAs), *control* instructions for when to make certain proteins, and instructions for making *transfer RNAs* (tRNA) and *ribosomal RNAs* (rRNA), which are important constituents of the machinery that assembles proteins. DNA contains no other instructions. Of the three kinds of instructions, this discussion concerns only the first one.

Recall further that proteins (or their component peptide chains) consist of linear sequences of amino acids. The linearity of amino acids in proteins and of bases in DNA is no coincidence. They are directly related. The sequence of bases in DNA constitutes the instructions for the sequence in which amino acids are to be assembled into a given protein.

There are a number of conceivable ways by which the sequence of bases in DNA could express instructions for the assembly of amino acids. For instance, we could imagine that base A refers to one kind of amino acid, G to another, and so forth.

A sequence of bases in DNA, such as CGACCATCAACT, would then have this meaning: assemble a peptide chain starting with the amino acid represented by C, add the amino acid represented by G, then the one represented by A, and so forth. Obviously, such *a ringlet code will* not work because the four bases could give assembly instructions for only four kinds of amino acids.

Somewhat more complex and versatile than a singlet code is *a doublet code,* in which the pairs of letters AA, AG, AC, . . . , TT represent the instructions for the assembly of amino acids. In this code the above sequence, CGACCA … , would state: start with the amino acid represented by CG, add the amino acid represented by AC, and so forth. Such a doublet code consists of 16 (4 × 4) different letter combinations and, hence, is still insufficient for directing the assembly of the twenty different amino acids found in protein.

One further increase in the number of letters designating the amino acids constitutes *a triplet code.* Such a code consists of the combinations AAA, AAG, AAC, … , TTT and comprises 64 (4 × 4 × 4) different combinations. This is more than enough to represent all twenty amino acids and is, in fact, the code used in nature .

The triplets of letters of the genetic code, which specify the amino acids, are called *codons.* In table elsewhere in this chapter the codons of DNA are listed for all twenty amino acids. There are many synonyms among the codons, as would be expected because the number of codons far exceeds the number of amino acids. The code also includes two *initiator* and three *terminator codons* that signal the beginning and the end of a genetic message. Using the table as a dictionary, we can now translate the above genetic message -CGACCATCAACT-into the corresponding peptide chain:

CGA CCA TCA ACT
| | |
Alanine-glycine-serine

The final codon of the message, ACT, is the signal for terminating the assembly of the peptide chain.

The example just given points to an interesting correspondence between the genetic code and writing in western culture. Both are linear in their representation of information. Both use letters and combine them into words. The letters of the genetic code are the bases and the words are the codons.

There are four bases and all of the codons are three letters long. In contrast, writing in English requires 26 letters and English words are of variable lengths. If we think of the codons as words, we may compare a sequence of codons, which gives instructions for the assembly of one protein molecule and is called a gene, to an individual article in a reference book. A DNA molecule, which carries many genes or many articles, may then be thought of as a book.

Table 1.2: Combinations of Bases in Singlet, Doublet, and Triplet Codes.

Singlet code (4 words)	*Doublet code (16 words)*				*Triplet code (64 words)*			
A	AA	AG	AC	AT	AAA	AAG	AAC	AAT
G	GA	GG	GC	GT	AGA	AGG	AGC	AGT
C	CA	CG	CC	CT	ACA	ACG	ACC	ACT
T	TA	TG	TC	TT	ATA	ATG	ATC	ATT
					GAA	GAG	GAC	GAT
					GGA	GGG	GGC	GGT
					GCA	GCG	GCC	GCT
					GTA	GTG	GTC	GTT
					CAA	CAG	CAC	CAT
					CGA	CGG	CGC	CGT
					CCA	CCG	CCC	CCT
					CTA	CTG	CTC	CTT
					TAA	TAG	TAC	TAT
					TGA	TGG	TGC	TGT
					TCA	TCG	TCC	TCT
					TTA	TTG	TTC	TTT

And all of the DNA molecules present in a cell, which together carry the totality of genetic information of an organism, correspond to the volumes making up an encyclopedia. All cells carry a full set of genetic information, at least when they first form, much as every library possesses an encyclopedia.

To conclude this section, consider how much genetic information is present in living cells. The DNA of a bacterium is typically 1 mm long and contains about 1 million codons. This is comparable to the number of words in five 500-page novels with 400 words per page.

It also is approximately equal to the words contained in one volume of a large encyclopedia. In contrast, the 23 DNA molecules (chromosomes) of a human cell have a total length of about 1 meter and carry approximately 1 billion codons.

This is roughly equal to the word content of 50 encyclopedias with 20 volumes each. This enormous difference in the quantity of genetic

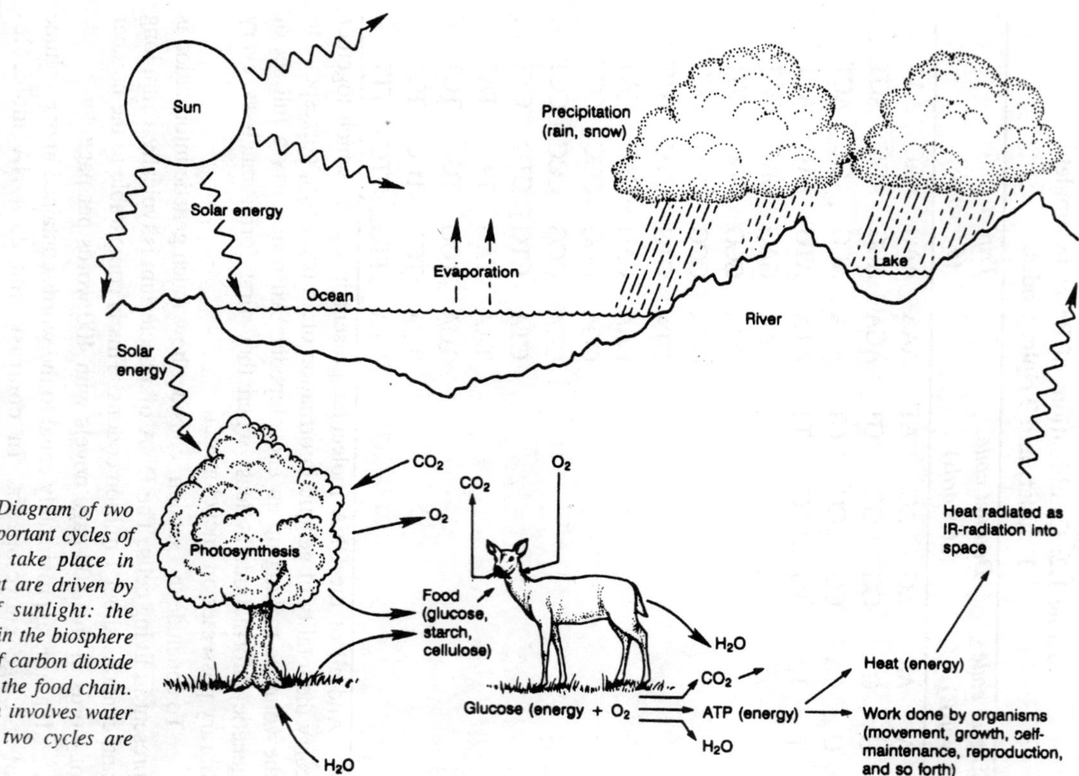

Figure 1.20 : Diagram of two of the many important cycles of materials that take place in nature and that are driven by the energy of sunlight: the cycle of water in the biosphere and the cycle of carbon dioxide and oxygen in the food chain. The food chain involves water also; thus the two cycles are interlinked.

information in bacteria and in humans is yet another indication of the range in diversity and complexity that has resulted in the course of 4 billion years of biological evolution.

This Introduction covers a wide range of biological topics in preparation for the chapters to come. Physical and biological evolution were compared, with emphasis on important similarities as well as differences between the two. The cell was introduced as the fundamental unit of biological organisation.

Prokaryotic and eukaryotic cells were distinguished and shown to represent the most profound demarcation in the biological world today. This demarcation is more profound than that existing between protists, fungi, plants, and animals, all of which are eukaryotes and evolved from prokaryotes, the most ancient living cells on Earth.

Biological evolution does not take place on the level of individual cells or organisms, but on the level of populations or species. This evolution proceeds by random genetic mutations together with a selection process that is the result of competition among the members of all species for available raw materials, energy, and living space. Because of this competition, species do not evolve in isolation, independently of each other.

They evolve together, strongly affecting each other's success in the struggle for survival. The stage upon which this evolution takes place is the physical environment of land, sea, and air, all of which evolve as well. Together, these evolving biological and physical components make up the Earth's ecosystem. The energy that drives the ecosystem is the energy of sunlight.

The range of these topics reflects the fact that the biological world may be viewed and studied on many different levels-from the level of molecules to that of the ecosystem, from the microscopic to the global. Every level has its importance and validity, and to understand biological evolution requires an awareness and appreciation of each of them.

Here the emphasis will be mainly on the level of species and ecosystems. Throughout these discussions we shall find that biological evolution on any level—molecular, cellular, organismic, species, ecosystem -invariably affects the global environment, often in profound and far-reaching ways.

2

SOLAR SYSTEM

Near the inner edge of the Orion Arm and approximately 30,000 LY from the center of the Milky Way, the Sun speeds silently through interstellar space at nearly 1 million kilometers per hour. Opposing gravitational and centrifugal forces keep it firmly locked into a nearly circular orbit around the massive, central bulge of the Galaxy, as noted in other chapter of this book.

Every 250 million years the Sun completes one circuit. It has done so since its birth some 4.5 billion years ago and, barring some near encounter or collision with another star, will do so in the future.

From a distance of a few light-years or more the Sun appears to be just another star among the 150 billion populating the Milky Way. However, when viewed from nearby it is a very special star. It is the gravitational hub of a planetary system that includes Earth—our home and shelter in the vastness of space-and, through its steady outpouring of radiation, provides the energy that sustains terrestrial life.

It is, therefore, not surprising that since the dawn of history humans have been fascinated by the Solar System. Our ancestors observed the daily rising and setting of the Sun, the waxing and waning of the Moon, and the wandering of the planets across the firmament.

Periodically they were startled and frightened by the appearance of a comet or the falling of meteorites, and they ascribed plagues and other calamities to these events. In their myths and religions they deified the Sun and the planets and they explained their motions by ingenious and, sometimes, fanciful theories.

Today we are still fascinated by the Solar System and employ science to probe its secrets. We use telescopes to observe the Sun, the planets, and the lesser bodies; we bounce radar waves off their surfaces; we send space probes to them; and in 1969, we first set foot on the Moon. As a result of the space missions, we have extensive photographic records of the surface features of the planets and their satellites, excepting only Neptune and Pluto. We know about their magnetic fields. We possess Moon rocks.

We have on-site analyses of the surface compositions of Venus and Mars. We are beginning to understand the structures of the planets and their satellites. And, with the aid of high-speed computers, we are developing new theories of the origin and evolution of the Solar System.

The goal of this chapter and the next is to describe the orbital distribution and structures of the member bodies of the Solar System, and to make you familiar with some of the current theoretical efforts concerning its origin and evolution. We shall begin with a brief summary of the historical background to our contemporary perspective.

EMERGENCE OF THE HELIOCENTRIC VIEW OF THE SOLAR SYSTEM

Most ancient scholars thought that the Earth was the center of the Universe, with all other celestial objects revolving around it. However, an early exception was Aristarchus of Samos (c. 310-230 B.C.), who advocated a heliocentric Solar System. The ancients clearly distinguished between the *wandering stars* or *planets* and the *fixed stars*.

The planets included the Sun, the Moon, Mercury, Venus, Mars, Jupiter, and Saturn, all of which, in the course of days to weeks, can be seen to wander across the firmament. The fixed stars were the objects still called stars today and that do not appear to partake in the wandering motions. Today we know that the planets and fixed stars differ not only in their apparent motions across the firmament, but are fundamentally different objects.

The planets belong to the Solar System and shine by reflecting sunlight. In contrast, the fixed stars (that is, those in our galaxy) lie at distances from a few to tens of thousands of light-years, and they shine by radiating energy released by nuclear reactions in their interiors.

The commonly held view among the ancients was that the seven planets as well as the fixed stars are carried on crystalline spheres around the Earth once every day and that the friction between the spheres gives rise to the "music of the spheres." This *geocentric* or *Earth-centered* view was formalised in the second century A.D. by the great Egyptian astronomer and mathematician, Ptolemy, in his book the

Almagest (derived from the Arabic, meaning “The Great Treatise”). He stated that all the planets move along small circular paths, called epicycles, whose centers in turn move along larger circles around the stationary Earth. This system allowed him to approximate the motions of the planets across the sky, including the retrograde motions of the outer three.

Copernicus and Kepler

Ptolemy’s geocentric view prevailed for more than 1300 years, until the early part of the sixteenth century. By then a new intellectual spirit had taken hold in Europe and many traditional teachings came under critical scrutiny. The mood was one of adventure and of looking beyond what was known or had been done before.

Explorers sailed the seven seas and merchants expanded their trade to the far corners of the globe. This new adventurism also spread to astronomy. Growing discrepancies were noticed between the Ptolemaic predictions of the positions of the planets and their true positions in the sky, and a few astronomers dared to acknowledge that something fundamental was wrong with the geocentric model.

The first major step in resolving the problem was taken by the Polish astronomer Nicolaus Copernicus, who in his celebrated work *De revolutionibus orbium coelestium* (“On the Revolutions of the Celestial Spheres”) proposed a new and much simpler model of the Solar System. He placed the Sun at the center of the system and let the *six* known planets, including Earth, move in circular orbits around it.

The Moon lost its status as a planet and became a satellite of Earth. This is the *Sun-centered* or *heliocentric* model of the Solar System. Although Copernicus’ heliocentric model came closer to reality than the geocentric model, it still did not quite correctly predict the motions of the planets.

This became particularly evident in the case of the planet Mars, whose positions were very accurately measured by the Danish astronomer Tycho Brahe over a period of some 20 years. Brahe attempted to resolve the disagreement between theory and observations, but without success.

Brahe’s assistant, the German mathematician Johannes Kepler, succeeded with the task. Kepler found that he could achieve agreement if he made two assumptions: One, the planets do not move in circular orbits, as proposed by Copernicus, but in elliptical orbits; and, two, the Sun lies not at the center of the orbits, but at one of the foci of the ellipses. He also noted that the planets travel faster when they are near the Sun and slower when they are more distant. Finally, he discovered a simple relationship between the planets' distances from the Sun and their orbital periods.

Kepler summarised these results in his famous ***three laws of planetary motion,*** which to this day form the basis of our understanding of celestial orbital motions. They apply to the motions of planets around the Sun, as well as to the motions of satellites, including artificial ones, around planets. They also apply to the motions of double stars and double galaxies around each other.

These accomplishments by Copernicus, Brahe, and Kepler contributed greatly to the intellectual optimism of their times. They had proved that it was possible to gain valuable insights into natural phenomena by making careful observations and measurements and by using them in the construction of new theoretical models.

No longer was it necessary to accept ancient dogmas on faith. We must rank this development -reliance on observation and experimentation combined with theoretical reasoning-among the great revolutionary breakthroughs in the intellectual history of humankind. It was, in fact, the beginning of modern science.

Galileo and Newton

The heliocentric model of the Solar System was met initially with much opposition, as Galileo's trial by the Inquisition amply demonstrates. Only in the succeeding centuries, after a series of remarkable discoveries confirmed the model's validity, did it become universally accepted.

Many scientists contributed to this success, but two of them stand out particularly. They are the two giants of seventeenth century science, Galileo Galilei and Isaac Newton. Galileo was the first to observe celestial bodies through the then newly invented telescope. He noticed that Venus has phases just like the Moon and correctly explained them as being due to that planet's motion around the Sun, during which increasing and decreasing portions of its illuminated side face toward us.

When he trained his telescope toward Jupiter, he saw four small bright objects orbiting around that planet with periods of between two and seventeen days. They are Jupiter's four largest satellites, which today we call the Galilean satellite—slo, Europa, Ganymede, and Callisto. Galileo fully realised the importance of this latter discovery.

He had found a new satellite system - a miniature version of the Solar System. Apparently, systems in which one or several objects revolve around a larger central object are common in nature, with the

Earth-Moon system being yet another example. Perhaps the most spectacular sight for Galileo was the surface of the Moon.

There he saw mountains, craters, valleys, and large dark areas, which he thought were seas. Obviously, that body is not the perfectly round and smooth sphere envisioned by the ancients. In many ways it resembles the Earth. Nor is the Sun a perfect body.

On its surface, Galileo found dark spots, which today we call sunspots. These blemishes had been observed before, but they were dismissed as being either illusions in the Earth's atmosphere or shadows cast by intervening planets. From the motions of the sunspots across the solar disk, Galileo deduced that our star rotates with a period of slightly less than one month.

Galileo described these and other telescopic discoveries in his book *Sidereus nuncius* ("The Starry Messenger").

Besides his astronomical discoveries, Galileo made numerous other major contributions to the development of modern science. In particular, through his experiments and mathematical analysis of the acceleration of material bodies, he paved the way for the work of Isaac Newton.

Newton was convinced that a fundamental principle governs the motions of all objects and he set out to find it. This eventually led him to realize that whenever a body experiences *a change in motion*—when it is either speeded up or slowed down or when the direction of its motion is changed a force is acting on it. He expressed his idea in three laws, which he published under the title *Philosophiae naturalis principia mathematica,* known as the *Principia.* With these three laws Newton laid the foundation of modern physics.

In the course of his work on the laws of motion, Newton wondered whether the force responsible for the falling of objects on Earth might not also be the force that keeps the Moon in orbit about the Earth, Legend has it that this idea first occurred to him while he was watching an apple fall from a tree.

He thought that, perhaps, the same force is also responsible for the planets' motions around the Sun as well as the Galilean satellites' motions around Jupiter. To test these hypotheses, he substituted the orbital data of the planets and satellites into Kepler's third law.

He found that all of the motions could be explained, including the motions of falling objects on Earth, if he assumed that the force (F) varies with the inverse square of the distance *(R)* separating the bodies: $F \propto 1/R^2$. He found further that the force is also directly proportional to the product of the masses of the bodies (for example, the masses of the Earth and the Moon, the Sun and a planet, M_1 and M_2): $F \propto M_1 M_2$. *By* combining the two proportionalities, Newton derived a mathematical expression for the *universal force of gravity:*

$$F \propto \frac{M_1 M_2}{R^2}$$

Universal means that this force applies not only to the effect of the Earth on falling apples and the Moon or of the Sun on the planets, but to the attraction between all matter and energy in the Universe. With this discovery, Newton demonstrated for the first time that there exists but one "design" in nature and that the scientific method allows us to discover it.

MODERN PERSPECTIVE OF THE SOLAR SYSTEM

During the three centuries following the pioneering work of Copernicus, Brahe, Kepler, Galileo, and Newton, knowledge of the Solar System grew at a steadily accelerating pace.

Three New Planets

In 1781, the German-English astronomer and musician Sir William Herschel noted, in the course of a systematic survey of the sky, that the image of a faint star was larger than is typical for stars. He observed the object for several weeks and found that it moved relative to the fixed background stars.

At first he thought he had discovered a comet. Several months later, with more positional data on hand, the object's orbit was computed. It turned out not to be the elongated orbit of most comets (more about them later), but a nearly circular path around the Sun at a distance of about nineteen times that of the Earth.

This is a distance greater than Saturn's and, judging from the object's brightness, meant that it was much bigger than a comet. Herschel had discovered a new planet. He named it *Georgium Sidus* (meaning "Star of George") in honor of George III, England's reigning king, but the name was later changed by J. E. Bode (see below) to *Uranus,* after the Greek god of the heavens. Incidentally, Herschel was not the first to sight Uranus.

The planet had been plotted on charts of the sky on at least seventeen previous occasions since the year 1690 without recognition of its true nature. By 1820, it had become clear that the orbit of Uranus is not exactly an ellipse. Over the years, its path in the sky deviated more and more from that predicted by theory, even after the perturbing effects of Jupiter and Saturn were taken into account. It appeared that an unknown and still more distant planet was pulling on Uranus. The question was where in the sky to find the unknown planet. Tentative positions were computed in 1841 by John Couch Adams, an undergraduate at Cambridge University in England, and, four years later, by Urbain Jean Joseph Leverrier in France.

A search was conducted for the unknown planet at the Cambridge and Berlin observatories. It was finally discovered in 1846 by Johann Gottfried Galle, a young German astronomer at the Berlin Observatory, within one degree of the position predicted by Leverrier. It was christened *Neptune,* in honor of the Roman god of the sea.

As in the case of Uranus, Neptune seems to have been sighted before it was identified as the Sun's eighth planet. Galileo's notebook shows that on several occasions in 1613 he had seen in the vicinity of Jupiter what he thought was a faint star that moved, and recent calculations indicate that it may well have been Neptune.

Early in the twentieth century, it became evident that Neptune alone could not account for the full perturbation that was observed in Uranus's orbit. There had to be yet another planet—*Planet X,* as it became known—that exerted a gravitational tug. Two American astronomers, William Henry Pickering and Percival Lowell, estimated that Planet X should lie somewhere in the constellation Gemini and that its mass should be between that of the Earth and Neptune.

An intense search ensued. Lowell looked for the planet from 1906 until his death, at the observatory he built in Arizona for the purpose of studying the planets. Success finally came in 1930 to Clyde Tombaugh, a 22-year-old Kansas farm boy and assistant at the Lowell Observatory. For many months he photographed sections of the night sky and compared photographic plates taken a few days apart.

His aim was to see if any of the stellar images showed displacement and, thereby, revealed itself as the missing planet. The work was tedious and time consuming. Then, on February 18, 1930, Tombaugh noted that the image of what he thought was a faint star had moved the expected amount on photographs taken six nights apart earlier in January. He had discovered the Sun's ninth planet.

It lay within 6 degrees of the position predicted by Lowell. The new planet was called *Pluto,* after the Greek god of the underworld. It was exceedingly faint and not sighted visually until 1950.

There is an ironic twist to Pluto's discovery that became evident only 48 years later. In 1978, James Christy, an astronomer at the U.S.

Naval Observatory, noted that on high-quality photographs Pluto's image has a bump and that this bump moves systematically around the planet. The bump turned out to be a satellite. Its orbit is so close to Pluto that on photographs the two bodies are not fully resolved, but merge into a single bumpy image.

The satellite has a period of 6.4 days, the same as Pluto's rotation period, and it circles Pluto at a distance of about 20,000 km (equal to three earth radii). Christy named the satellite *Charon,* for the boatman of Greek mythology who ferries the souls of the dead to Hades, the underworld and Pluto's domain. Charon's orbital period and distance imply-through Kepler's third law-that Pluto's mass is only 0.002 earth masses, which is considerably less than the mass assumed by Lowell in making his prediction of Pluto's position.

It is, in fact, such a small mass that it cannot possibly be responsible for the observed residual perturbation in Uranus's orbit. Therefore, Pluto's discovery on the basis of Lowell's prediction was pure coincidence. It was, however, a lucky coincidence: without it we might still not be aware of Pluto's existence.

If Pluto is not Planet X, perhaps there exists a larger and as yet undiscovered planet at a still greater distance from the Sun that is the true source of the residual perturbation of Uranus's orbit. The case for this hypothesis is strengthened by small and unaccounted for discrepencies in the orbits of the other planets, from Jupiter to Neptune.

In fact, the case for a tenth planet is so strong that astronomers in the United States and Europe are now actively searching for it. They do not presently know where exactly in the sky it is located, if, in fact, it does exist; but they estimate that it has a mass between two and five earth masses and an orbit beyond Pluto's, at about 50 to 100 astronomical units from the Sun.

The Planets' Orbital Characteristics

Before continuing with the description of the Sun's planetary system, a summary of the planets' orbital characteristics is in order:

1. All planets revolve around the Sun in nearly the same plane. Relative to the Earth's orbital plane-also known as the *plane of the ecliptic—the* other planets' orbits are inclined by only three degrees or less. The only exceptions are Mercury and Pluto, whose orbits are inclined by 7 and 17 degrees, respectively.
2. The orbits of all but two of the planets are nearly circular. Again, the two exceptions are Mercury and Pluto, whose orbits

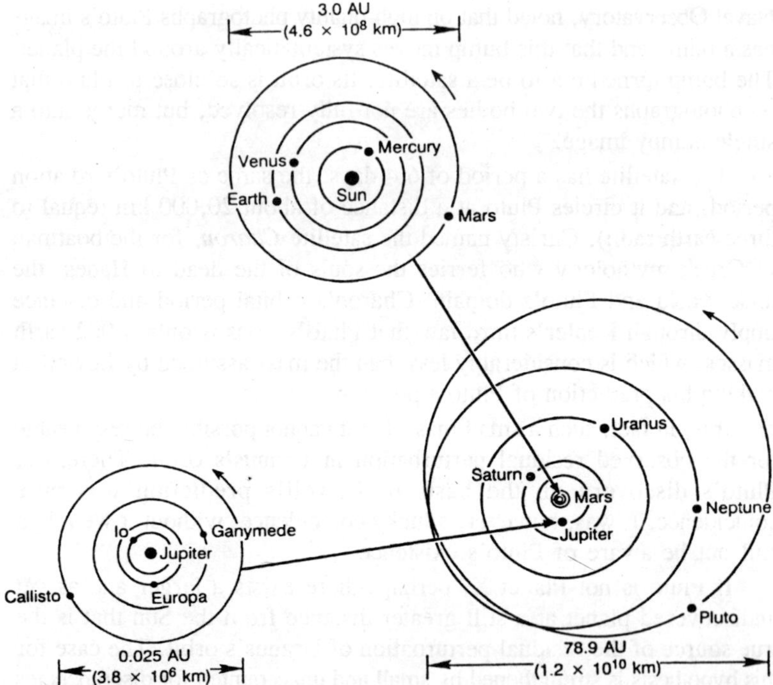

Figure 2.1 : The relative sizes of the orbits of the planets, including those of Jupiter's Galilean satellites. The Sun is not drawn to scale (it is 80 times smaller than Mercury's orbit).

have eccentricities of 0.21 and 0.25, respectively.

3. All planets orbit the Sun in the same direction, namely from west to east. This direction is called *prograde,* in distinction to *retrograde,* which refers to the opposite direction.
4. With two exceptions, the planets rotate in the same direction in which they orbit. That is why, for example, on Earth the Sun rises in the east and sets in the west. Only Venus and Uranus are different. Both rotate in the retrograde direction. However, while Venus' equator is inclined by only 2° to its orbit, Uranus' equator is tilted almost perpendicularly to its orbit.
5. The planets orbital distances from the Sun are so regular that they can be approximated by a very simple numerical "law," discovered in 1781 by Johann Daniel Titius of the University of Wittenberg in Germany. Titius found that taking the sequence of nine numbers 0, 3, 6, 12,... , 384, adding 4 to each, and

dividing the sums by 10, yields approximately the distances from the Sun to Mercury, Venus, and so forth in astronomical units (with the single exception of Neptune), as shown in the table on the following page. This law is named after its discoverer and after Johann Elert Bode who, as director of the Berlin Observatory, published and popularised it.

Planet	*Distance from the sun according to the Bode-Titius Law (AU)*	*Actual average distance from the Sun (AU)*
Mercury	(0 + 4)/10 = 0.4	0.39
Venus	(3 + 4)/10 = 0.7	0.72
Earth	(6 + 4)/10 = 1.0	1.00
Mars	(12 + 4)/10 = 1.6	1.52
Asteroids	(24 + 4)/10 = 2.8	2.70
Jupiter	(48 + 4)/10 = 5.2	5.20
Saturn	(96 + 4)/10 = 10.0	9.55
Uranus	(192 + 4)/10 = 19.6	19.2
Neptune		30.1
Pluto	(384 + 4)/10 = 38.8	39.4

Asteroids

At the time of the discovery of the Bode-Titius law, the space between Mars and Jupiter was marked by a glaring gap. The law predicted a planet at a distance of 2.8 AU, but none was known to exist. The problem was solved 20 years later in 1801, when a small body approximately one-third the size of the Moon was sighted orbiting the Sun at almost the exact distance predicted.

It was named *Ceres,* after the Roman goddess of agriculture. Soon other similar bodies were found, most of them in the gap between Mars and Jupiter, and today we know of thousands of them. Most are less than one kilometer across and their total mass is no more than one-tenth the mass of our Moon. These bodies are known as the *asteroids* or *minor planets.*

The word *asteroid* comes from the Greek and means "starlike," although these bodies are not like stars at all. Most of them are rocky objects, and all but the largest ones are irregularly shaped and heavily marked by collisions they have suffered over the eons. Some appear to be double or multiple bodies and are thought to be the broken-up pieces of formerly larger bodies. All asteroids are believed to be leftover

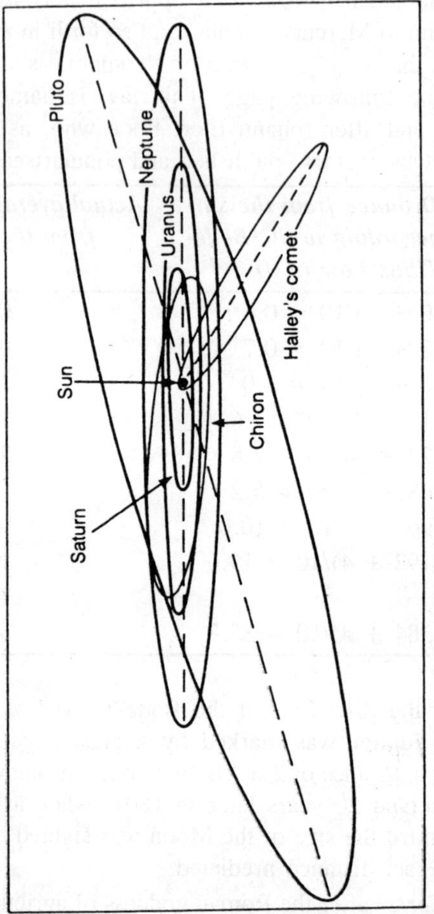

Figure 2.2 : Diagram of the orbits of Saturn, Uranus, Neptune, and Pluto; the asteroid 2060 Chiron; and Halley's comet; as viewed from 5° above the ecliptic. Pluto, Chiron, and Halley's comet have noticeably elongated and inclined orbits.

debris from the formation of the Solar System. A later discussion will show why they never managed to condense into one single planetary mass.

Most asteroids orbit the Sun in a broad belt, the *Asteroid Belt*, at an average distance from the Sun of 2.7 AU. However, not all asteroids are so confined. Some have very elliptical orbits that extend inward past the orbits of Mars and Earth.

They are the *Amor* (Mars-crossing) asteroids and *Apollo* (Earth-crossing) asteroids. Asteroid Icarus, which is a chunk of rock 700 m across, has an orbit that carries it to within 0.19 AU of the Sun (in comparison, Mercury orbits the Sun at roughly twice that distance) and outward past the orbit of Mars.

Other asteroids stray outward to beyond the orbits of Saturn and Uranus as, for example, Chiron (or Asteroid 2060), discovered in 1977. Still others, the *Trojan Asteroids* (most of which are thought to be icy rather than rocky objects), are firmly locked into orbits with the same period as Jupiter's, with one group traveling 60 degrees ahead and the other group 60 degrees behind Jupiter, as seen from the Sun's position.

Many of the smaller satellites of the planets are thought to be asteroids that were captured by the planets. These small satellites have masses, sizes, and compositions like asteroids. For example, the two moons of Mars, Phobos and Deimos, have dimensions of only 14 × 10 km and 8 × 6 km, respectively.

They are irregularly shaped and photographs reveal surfaces marred by many large and small impact craters, indicating countless collisions in the past with other rocky debris in interplanetary space. Eleven of Jupiter's 16 known satellites are less than 100 km across and the outer four of them are in retrograde orbits.

Twelve of Saturn's 23 known satellites are also quite small (100 km in size or smaller), and so are 10 of Uranus's 15 known satellites. No doubt, as astronomers continue their explorations of the outer Solar System, they will discover many additional small asteroid-like satellites circling the giant planets.

Comets

Far beyond the planets, in the dark and cold outer fringes of the Solar System, many billions of small, primordial bodies of rock and ice are suspected to circle the Sun. They constitute the *Oort Cloud* of dormant cometary bodies, which, like the asteroids, are thought to be leftover debris from the formation of the Solar System. This cloud is believed to extend up to 100,000 AU outward from the Sun, which is more than a third of the way to the Alpha Centauri system (the nearest stars, at a distance of 270,000 AU).

The existence of this cloud of cometary bodies was first proposed in 1950 by the Dutch astronomer Jan Oort, in order to explain the frequency of appearance of comets in the inner part of the Solar System. According to Oort, the occasional passage of nearby stars perturbs the orbits of some of these bodies and deflects them toward the inner part of the Solar System.

There, bombardment by radiation and charged particles from the Sun brings them to "life." Within about 3 AU of the Sun their ices start vaporizing, and gas and dust particles stream outward to form an envelope -the *coma* and *hydrogen cloud-around* the central body, the *nucleus*. Pressure exerted by the solar radiation and solar wind drives the gases and dust many millions of kilometers through interplanetary space to form the cometary *tail*. Due to the way it is formed, the tail always points away from the Sun. The main features of the components of comets are summarised in table elsewhere in this chpater.

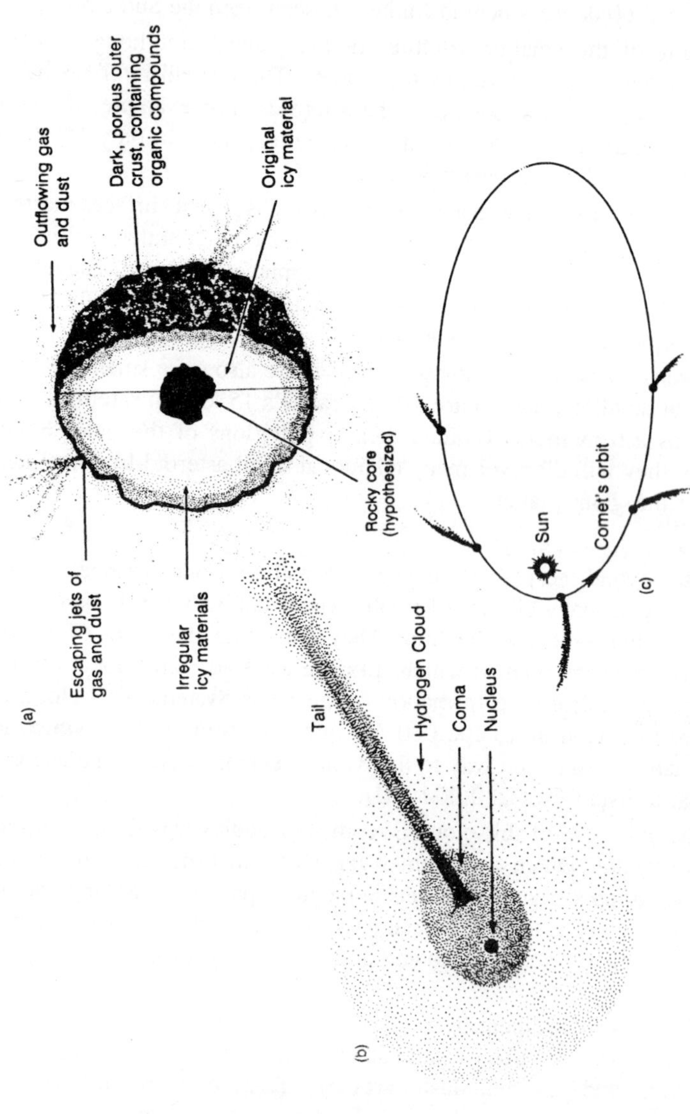

Figure 2.3 : The structure of comets. (a) Model of comet nucleus; (b) Diagram of major components of comets; (c) Shape and growth of tail of typical comet as it passes near the Sun.

Table 2.1: Features of Components of Comets.

Component	*Composition*	*Size*
Nucleus: central solid body of comet	rocks and ices of water (H20), methane (CH4), and ammonia (NH3), source of all the dust and gas of the comet's coma, hydrogen cloud, and tail	100 m-25 km irregularly shaped; *mass:* 10^9–10^{17} kg (10^6-10^{14} tons)
Coma: envelope of gas and dust surrounding nucleus	gas: neutral atoms and molecules rich in hydrogen, carbon, oxygen, nitrogen, sulfur, and various metals (H, C, O, S, OH, CH, CN, CO, CS, NH, NH_2, HCN, CH3CN, Na, Fe, K, Ca, V, Cr, Mn, Co, Ni, Cu) *dust:* microscopic silicate minerals (similar to terrestrial sandy dust but less compact)	10^5-10^6 km (approx. the size of the Sun)
Hydrogen Cloud: enormous nearly envelope	atomic hydrogen, derived from the disintegration (by sunlight) of OH and other hydrogen-rich radicals of the coma	up to many million km (much larger than the Sun) spherical
Tail: elongated structure of dust and gas	*dust:* primarily silicate minerals gas: charged particles (e^-, CO^+, CO_2 , H_2O^+, OH^+, CH^+, CN^+, N_2 , C^+, Ca^+)	*length: up* to 1 AU

Sunlight reflected from the gas and dust particles produces the image that we commonly have of comets—the long, sweeping tail, extending out from the bright, globular coma. We obtain this image of comets mostly from photographs. With the unaided eye comets can usually be seen only as inconspicuous, faintly glowing objects, right before sunrise or after sunset.

However, occasionally comets do become spectacular. Then they arouse people's interest and imagination, and even strike awe and fear in some of them. That was the effect Comet Halley had during its apparition in 1910. About four comets visit the inner part of the Solar System every year.

Most of them come from far beyond the orbit of Pluto, approach the Sun in highly elongated orbits, whip around it, and then recede again. Hundreds of thousands of years may go by before they return once again. There are, however, a number of comets that by near encounters with a planet were placed into orbits closer to the Sun.

If their orbital periods are less than 200 years (corresponding to about 35 AU for the length of the semimajor axis of their orbits), they are called *short period comets*. If their periods are greater than 200 years, they are called *long period comets*. Comet Halley (named after the British astronomer and mathematician Edmond Halley is a short-period comet.

It completes an orbit every 75 to 80 years, approaching the Sun to within 0.59 AU and receding to 35 AU. This comet has been sighted regularly since 240 B.C., when Chinese records first make mention of it. Its most recent approach to the inner part of the Solar System occurred in 1985-86, when it was extensively investigated and photographed by five unmanned spacecraft. Every time a comet passes through the inner part of the Solar System it loses some of its constituent gases, dust, and rocks.

After hundreds to thousands of such passages, nothing is left but a fragile rocky body that will eventually break up due to the Sun's tidal forces. The rocky rubble that results continues to travel along the comet's original orbit. In the course of time, the combined gravitational effects of the Sun and the planets disperse the cometary debris further and further until it isspread over the entire orbit. This debris is the source of shooting stars or meteors.

3

Origin of Planetary Bodies

In other chapter of this book we have primarily concerned ourselves with the orbital layout of the Solar System. In order to progress toward our goal of understanding the origin of the Sun and the planets, we must also learn about the physical makeup of the system's member bodies. The structure of the Sun was discussed in other chapter. Now let us look at other bodies in the Solar System and, in the last section of this chapter, at some of the current theories of its origin and evolution.

THREE FAMILIES OF BODIES IN THE SUN'S PLANETARY SYSTEM

The bodies circling the Sun and constituting the planetary system range from dust and pebble-sized debris to planets with radii of thousands to tens of thousands of kilometers. The lower part of this range includes the asteroids, comets, many (but not all) of the satellites of the planets, the particles composing the rings of Jupiter, Saturn, Uranus and Neptune, and the interplanetary dust and rock fragments that have resulted from collisions among the bodies.

Including among these *small bodies* of the Solar System those objects that have radii and masses up to approximately 1000 km and 1/1000 earth masses (a few times 10^{21} kg). Many of them have orbits that are highly elliptical and inclined to the ecliptic.

Together they form a family of bodies with a common history and with compositions that, on the average, vary systematically with distance from the Sun. They are believed to be the leftover debris from the

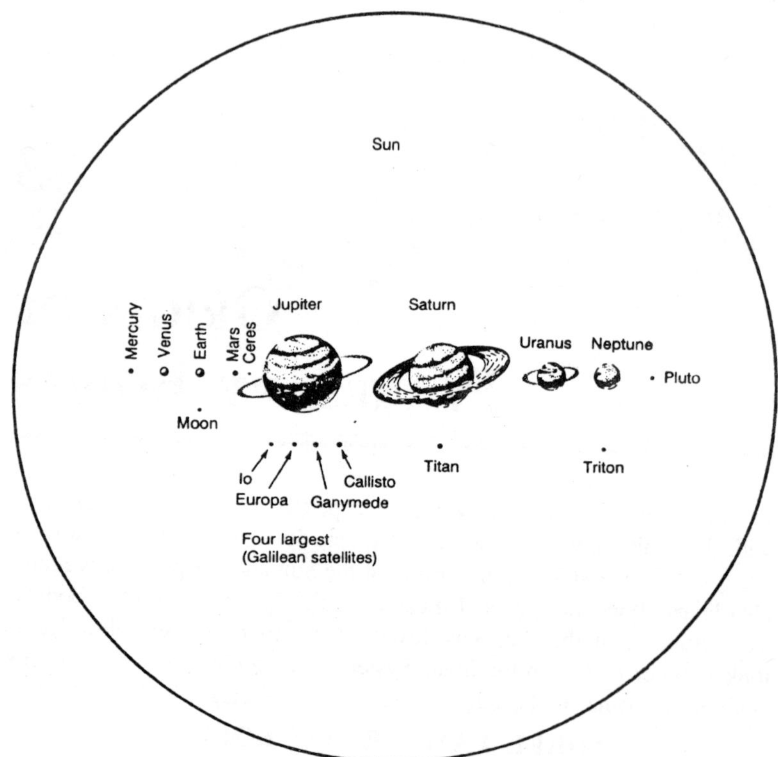

Figure 3.1: Relative sizes of the Sun, the planets, the largest of their satellites, and Ceres.

original material out of which the planets and their large satellites were assembled. In the theory of Solar System formation, such small bodies are referred to as *planetesimals.*

A second family of Solar System bodies includes the terrestrial planetsMercury, Venus, Earth, and Mars. In this book, the Moon, the Galilean satellites, Saturn's satellite Titan, and Neptune's satellite Triton are also included in this family. We refer to them as the *intermediate-sized planetary bodies.*

They have radii between roughly 1500 km and 6500 km; their masses range from about 1/200 to 1 earth mass; they have similar structures; and related physical processes shaped them. The orbits of most of them are nearly circular, with inclinations either close to the ecliptic or to the equatorial plane of the central body about which they revolve.

The major exceptions are Mercury, as already noted, and Triton.

Table 3.1 : Viewing Data and Orbital Characteristics of the Four Meteor Showers Whose Orbits.

Meteor shower	*Date of maximum display*	*Normal period of visibility*	*Constellation in which meteor tracks appear*	*Period of original comet (years)*	*Velocity relative to Earth (km/Sec)*	*Eccen-tricity of orbit*	*Inclination of orbit to ecliptic (degrees)*
Quadrantids	Jan 3	Jan 2-4	Bootes	7	43	0.71	75
Perseids	Aug 12	July 29-Aug 18	Perseus	110 (retrograde)	61	0.96	65
Leonids	Nov 16	Nov 14-19	Leo	33 (retrograde)	72	0.92	17
Geminids	Dec 13	Dec 8-15	Gemini	1.6	37	0.90	26

Triton is in a retrograde orbit that is inclined by 20° to Neptune's equator. As in the case of the small bodies of the Solar System, the compositions of the intermediate-sized planetary bodies also vary systematically with distance from the central body about which they revolve.

In the past, Pluto has been included in the family of intermediate-sized planetary bodies; but, as this family is defined here, Pluto is a marginal case. Pluto's size is comparable to Europa, the smallest of the Galilean satellites; yet its mass is only 1/460 earth masses, which places it somewhere between a large asteroid and a small intermediate-sized planetary body.

Furthermore, Pluto's orbit is quite elliptical and inclined, similar to the orbits of asteroids and comets. Hence in this book Pluto (as well as its companion, Charon) is regarded as a member of the family of small bodies of the Solar System, even though in the tables it will be listed among the planets.

Of course, despite the division made here, there is no really natural or absolute borderline between the families of small and intermediate-sized planetary bodies. Together these bodies constitute a continuum from microscopic dust through small irregularly shaped boulders, to large, roundish bodies of rocks, metals, and volatiles. The choice of the borderline between the two families is quite arbitrary and mainly one of convenience.

The third family of Solar System bodies are the *giant or Jovian planets* Jupiter, Saturn, Uranus, and Neptune. Their radii extend from roughly 25,000 to 70,000 km and their masses range from approximately 15 to more than 300 earth masses. Their structures are largely gaseous and liquid, and they have been formed by physical processes that were somewhat different from those of the smaller bodies.

Their sizes, structures, and origins place these giant planets somewhere between the terrestrial planets and a small star, though none of them is even close to generating energy by nuclear fusion. Unlike the intermediate-sized planetary and small bodies in the Solar System, they do, however, radiate more energy than they receive from the Sun by processes that are at present not fully understood. (Summaries of the physical characteristics of members of the three families of Solar System bodies are found in tables else where in this chapter).

SMALL BODIES OF THE SOLAR SYSTEM

The best way to study the small bodies of the Solar System would

Table 3.2: Physical Characteristics of Meteorites

		Dominant characteristics	*Density* *(gkm³)*	*Elemental composition*	*Comments*
Undifferentiated Meteorites	Ordinary chondrites	Fine-grained earthy matrix containing two distinct components: (1) hard irregularly shaped *mineral inclusions* with high concentrations of Ca, Al, Ti, and (2) round or oval *mineral chondrules*[b] with high concentrations of Fe, Mg, Si.	about 3.7	Solar composition, except volatiles are underrepresented.	*Ca-Al-Ti inclusions* in chondrites are the most refractory constitu ents found in meteorites, meaning they were at one time exposed to high temperatures (1500 K or higher) that vaporized and drove off the volatile compounds. *Fe-Mg-Si chondrules* were once molten droplets, probably created during high-speed collisions of planetesimals.
	Carbonaceous chondrites	Similar to ordinary chondrites including the presence of Ca-Al-Ti inclusions, but chondrules are fewer in number or absent, and H_2O (2 to 20% by mass) and *organic compounds (up* to 5%) are present.	2.2-3.6		Organic compounds of carbona ceous chondrites include amino acids and tarlike chains of hydro carbons. Among the amino acids, glycine, alanine, valine, proline, and glutamic acid are abundant, as well as amino acids not present in biologically synthesized protein. Both L- and D-amino acids are

(Table Contd.)

(*Table Contd.*)

		Dominant characteristics	*Density* (*gkm³*)	*Elemental composition*	*Comments*
					present, indicating an abiotic origin.
Differentiated Meteorites	Achondrites	Similar to terrestrial and lunar igneous rocks.	3.0-3.5	Composition very different from solar due to process of differentiation.	Differentiation is believed to have ated occurred in parent bodies (that is, the planetesimals) of solar or ordinary chondritic composition and with diameters exceeding several ten kilometers.
	Stony irons	Roughly equal proportions of silicate minerals and Fe-Ni metals.	5.5-6.0		
	Irons	About 90% Fe, 5-10% Ni.	7.5-8.0		

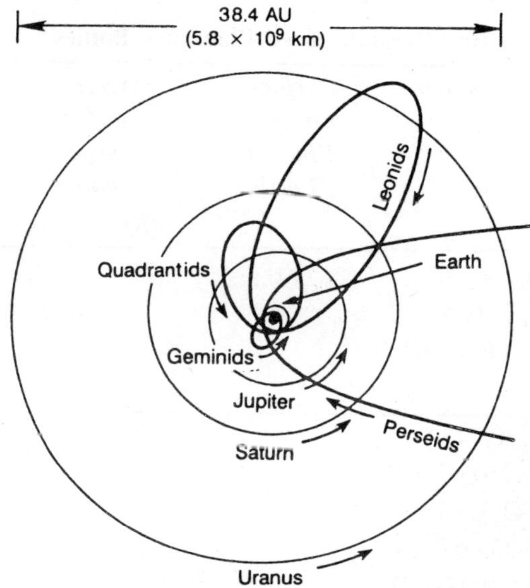

Figure 3.2 : Diagram of the orbits of four meteor showers. The shapes (that is, eccentricities) and sizes of the orbits are drawn to scale.

be to send space probes to the Asteroid Belt and comets and to collect samples. That is what the United States did with its *Apollo* missions to the Moon. Similar missions to the asteroids and comets are technically feasible, but would be so expensive that to date they have not been funded.

Until there is the opportunity to collect samples of asteroids and comets directly, scientists have to make do with the information they can obtain through remote sensing, flyby missions, and by collecting and analyzing whatever debris from these smaller bodies happens to collide with our planet and survives the fall to the ground.

Meteorites

As it turns out, many asteroids or fragments of asteroids, as well as the rocky remnants of past comets, are in elongated Earth-crossing orbits, and every year Earth collides with and sweeps up roughly ten thousand tons of this interplanetary rubble.

Most of these encounters are harmless, for the rubble is usually microscopic in size and burns up by friction as it passes through the Earth's upper atmosphere, briefly creating fiery trails across the sky at night. Such trails are popularly known as *shooting stars* and technically as *meteors*.

Table 3.3: Characteristics of the Atmospheres of Five Intermediate-Sized Planetary Bodies.

	Composition of gases (fractions by number)	*Surface pressure PPlanet/ PEarth*	*Average surface temperature (K)*	*(C)*	*Surface gravity (m/sec²)*
Venus	CO_2 (0.96) N_2 (0.035) H_2O (0.0001) SO_2 (0.0001)	90.	730	457	8.9
Earth	N_2 (0.77) O_2(0.21) H_2O (0.01) Ar (0.0093) CO_2 (0.00033)	1.	288	15	9.8
Mars	CO_2 (0.95) N_2 (0.027) Ar (0.016) O_2(0.0013) CO (0.0007) H_2O (0.0003)	0.007	218	- 55	3.7
Titan	N_2 (0.80-0.95) CH_4 Organic Molecules Smog Particles	1.6	93	-180	1.4
Triton	CH_4	0.0001	(?)	(?)	0.9

Note: O_2 is molecular oxygen, SO_2 is sulfur dioxide.

Some meteors arrive in groups of thousands regularly every year and can give rather spectacular displays. Viewing data and orbital characteristics of a few *meteor showers* are given in table and figure elsewhere in this chpater.

Occasionally, large bodies weighing several kilograms or even tons enter the atmosphere. Typically they travel at speeds of a few times 10 km/sec (about 10^5 km/hr) and friction with the air sometimes produces so much heat that they may become brighter than the full Moon. Often they are sheared apart into two or more fragments. The largest of them, weighing thousands of tons or more, create sonic booms that can be heard for many miles.

These large bodies do not burn up, but pass right through the atmosphere with very little deceleration. Most of these *meteorites* fall into the ocean, but about one-third of them do come down on land where they carve out craters that may be several meters to many tens of kilometers across.

Fortunately, impacts by meteorites are relatively rare today, and the probability of being struck by one is extremely low. Still, on every continent there exist a number of craters, such as the Barringer Crater near Winslow, Arizona, that give evidence of violent collisions between Earth and interplanetary bodies.

Another example of a collision of Earth with an interplanetary body is the violent explosion that shook the Tunguska region of Siberia on the morning of June 30, 1908. Apparently, a cometary fragment entered the Earth's atmosphere and exploded just above the ground. People as far as 750 kilometers away saw the event, and the energy released by the explosion scorched a 30-kilometer wide area and leveled trees like matchsticks.

Planetary scientists get greatly excited about new finds of meteorites because they are presently the only samples of asteroid and comet material. Because asteroids and comets are thought to be part of the original material from which the terrestrial planets and the cores of the Jovian planets were assembled, analysis of their chemical and structural makeup is expected to reveal much about the early history of the Solar System.

There exist two fundamentally distinct types of meteorites: meteorites that have changed very little since their formation and meteorites that have experienced major alterations. Meteorites that have changed very little are said to be *undifferentiated* or *primitive,* and they have compositions resembling that of the Sun, except for a paucity in the volatiles (such as the noble gases and compounds containing H, C, N, and O).

Among these meteorites astronomers recognize two subspecies -the *ordinary chondrites* and the *carbonaceous chondrites*. Meteorites that have experienced major alterations are called *differentiated* meteorites. Among them astronomers recognize three subspecies-the *achondrites*, the *stony-iron,* and the *iron* meteorites. The main features of these five kinds of meteorites are summarised in table elsewhere in this chapter.

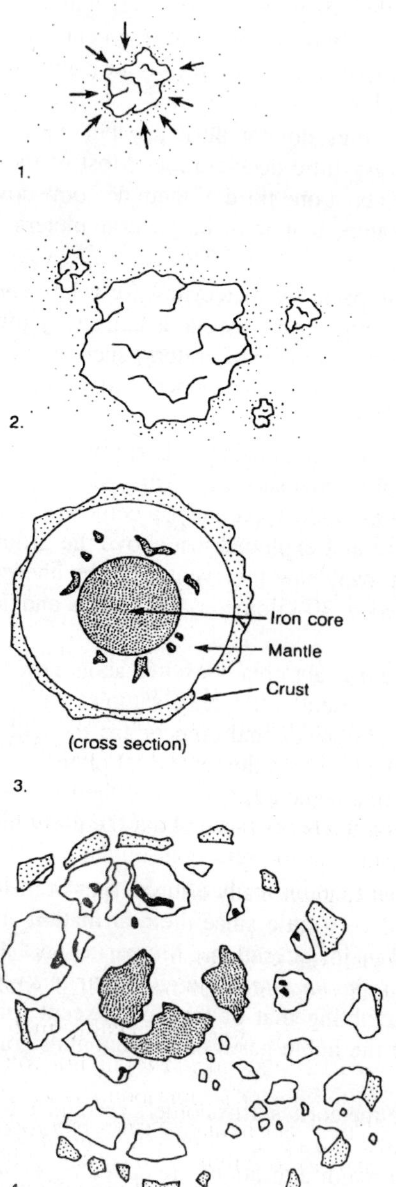

Figure 3.3: Four major stages in the formation history of stony, stony-iron, and iron meteorites.

Planetesimals

The composition of the meteorites indicates that they -or their parent bodies, the *planetesimals-formed* by the coalescence of fine- and coarse-grained mineral matter, metals, and gases of various kinds, including organic compounds.

This coalescence probably took place in the *protoplanetary disk,* which theory suggests surrounded the nascent Sun and initially was of the same chemical composition as the material that went into the making of the Sun.

Two different mechanisms are believed to have been at work in the formation of planetesimals. One mechanism produced the undifferentiated planetesimals, while the other mechanism produced the differentiated planetesimals:

1. Origin of the undifferentiated planetesimals (ordinary and carbonaceous *chondrites).* The undifferentiated planetesimals were bodies that never grew much larger than a few ten kilometers across, which was too small for internal melting and differentiation by gravitational forces.

However, these planetesimals acquired rather different compositions, depending on where in the protoplanetary disk they formed-compositions resembling those of ordinary chondrites in the inner part of the protoplanetary disk to compositions resembling those of carbonaceous chondrites in the middle and outer parts.

These differences in composition resulted from the differential heating the protoplanetary disk experienced, with temperatures ranging from above 1500 K near the young or still forming Sun to below 100 K hear the outer edge of the disk. The matter near the inner part of the disk, which was exposed to the highest temperatures, lost all but the most refractory minerals. Only the calcium-aluminum-titanium (Ca-Al-Ti) inclusions found in chondrites and a few other silicates survived there.

Matter more distant from the Sun, which was exposed to somewhat lower temperatures, was able to retain the less refractory minerals as well as some of the volatiles, including carbonaceous compounds. Finally, matter in the outer regions of the disk, which never became heated much at all, retained all elements (with the possible exception of the noble gases), including ices of water, ammonia, and methane.

2. Origin of the differentiated planetesimals (achondrites stony-iron, and iron meteorites). The bodies that gave rise to the differentiated meteorites grew initially by the same process that formed the undifferentiated planetesimals. However, in time they grew to much larger sizes and they became

heated to the point of partial melting. The densest of the elements -mostly iron and nickel -sank toward the centers of the bodies, while the lighter materials-the silicates and volatiles-rose to the outer layers. Thus, the bodies became differentiated into iron-nickel cores surrounded by mantles

whose compositions varied from mixtures of iron and stone to pure stone. Subsequent collisions or rotational instabilities broke the differentiated bodies into fragments and created the achondrites and stony-iron and iron meteorites. (The same mechanism of differentiation operated in all of the terrestrial planets and larger satellites.)

The variation in composition acquired by the planetesimals due to the temperature gradient in the protoplanetary disk is still in evidence today among the asteroids and comets. For example, asteroids near the inner edge of the Asteroid Belt (at distances of about 2 AU from the Sun) generally consist of rocks (comparable to ordinary chondrites), rocks intermixed with iron (stony-iron meteorites), and iron-nickel (iron meteorites).

The densities of the iron asteroids are 7.5 to 8.0 g/cm^3, while those of the rocky and stony-iron asteroids range from 3.0 to 6.0 g/cm^3, depending on the amount of iron they contain. Many of the asteroids in the middle of the belt (at about 3 AU from the Sun) are carbonaceous chondrites, with densities of 3 g/cm^3 or somewhat less, though there are still some stony and stony-iron as well as an occasional iron asteroid among them. In the outer belt and beyond (that is, beyond 4 AU), virtually all of the asteroids are of the carbonaceous type.

Many are reddish in colour, indicating increased admixtures of organic matter, including amino acids. Finally, the space beyond the planets is occupied by comets, which consist of rocks intermixed with ices of H_2O, NH_3, and CH_4 as well as many other volatiles. They have densities of about 1 to 2 g/cm^3.

Deviations from this general pattern occur. Objects of cometary origin can be found close to the Sun, and stony or stony-iron asteroids exist beyond the Asteroid Belt.

These exceptional cases are thought to be the result of orbital perturbations that the bodies experienced by past collisions among themselves, by gravitational interactions with Jupiter and the other planets, or, in the case of the comets, by gravitational interactions with passing stars.

What could have been the sources of heat that were responsible for the melting of the larger planetesimals? There is at present no sure answer to this question, though it is known that the gravitational energy released during the formation of the planetesimals was far too insignificant to have caused melting of bodies of only a few or even several hundred kilometers in size.

However, one source of heat that may have been very "important in the melting of at least some of the planetesimals was the decay of the radioactive isotope aluminum-26 (^{26}M) into magnesium-26 (^{26}Mg). The telltale sign of this decay is a slight excess of ^{26}Mg, relative to its abundance in solar material, in the Ca-Al-Ti inclusions of ordinary chondrites.

Presumably, the excess ^{26}Mg was not incorporated into the inclusions as magnesium, but as ^{26}A1, which subsequently decayed into ^{26}Mg and released energy. The ^{26}A1 must have come to the protoplanetary material within a few million years of being synthesized, as we can infer from this isotope's relatively short halflife of 720,000 years.

Had ^{26}A1 been synthesized much earlier, most of it would have decayed into ^{26}Mg and given off its energy before it ever became incorporated into planetesimals. The only known sources of ^{26}A1 are red giants, nøvae, and supernovae. We do not know which of these potential sources actually supplied the ^{26}A1 to the protosolar material.

However, Alastair G. W Cameron of Harvard and James W Truran of the University of Illinois made the interesting suggestion that if the 26A1 came from a supernova -perhaps one that exploded within a few dozen light-years from the interstellar gas cloud out of which the Solar System formed -the shock wave from the explosion may have been the trigger of the cloud's collapse.

Age of the Solar System

The decay products of long-lived unstable isotopes in meteorites allow us to deduce yet another important characteristic about the Solar System, namely its age. For example, potassium-40 (^{40}K) decays into argon-40 (^{40}Ar) with a halflife of 1.3×10^9 years. The amount of ^{40}Ar present in chondrites indicates that ^{40}K has been decaying in these meteorites for the past 4.5 to 4.6 billion years.

Because the chondrites have experienced no changes in their composition other than by radioactive decay of unstable isotopes, this must be the age of these meteorites. It is also regarded to be the age of the Sun and the planets, for it is believed that all of these bodies formed together out of the same interstellar cloud. In addition, similar ages are obtained from the decay products present in some lunar rocks as well as from theoretical model studies of the Sun. (In comparison, minerals and rocks from geologically active planets like the Earth, in which the composition is continually altered physically and chemically, give widely varying ages, from more than 4 billion years to very recent ages.)

INTERMEDIATE-SIZED PLANETARY BODIES: INTERNAL STRUCTURE

The family of intermediate-sized planetary bodies includes Mercury, Venus, Earth, Mars, the Moon, the Galilean satellites, Titan, and Triton. These bodies have similar sizes, masses, and other characteristics, as noted earlier and shown in tables elsewhere in this chapter. They also have similar, though not identical, structures.

The differences that do exist between them can be explained in terms of their masses, their different distances from the Sun or, in the case of the Galilean satellites, from Jupiter, and the kinds of materials that were available for their assembly at those distances. Let us examine the compositions and physical structures of these bodies, starting with the terrestrial planets.

The single most meaningful parameter concerning the composition of a planetary body is its density. A body made entirely of iron has a density of about 7.5 g/cm^3; one made mainly of silicate rocks has a density of about 3.0 to 3.5 g/cm^3; one made of rocks plus some admixture of volatiles has a density of between 2.0 and 3.0 g/cm^3; and one containing mostly ices has a density between 1.0 and 2.0 g/cm^3. (In comparison, basalt and granite, the major constituents of the Earth's rocky crust, have density ranges from 2.7 to 3.4 and 2.5 to 2.8 g/cm^3, respectively.)

Care must be taken in using density as an indicator of a celestial body's internal constitution. If a body's radius exceeds a few thousand kilometers, its self-gravity becomes so strong that even rocks and metals become appreciably compressed. In such cases it is not the actual density that is meaningful, but the body's *uncompressed* density, namely the density it would have if it were not compressed by self-gravity.

Terrestrial Planets

The distances from the Sun, radii, and actual and uncompressed densities of the terrestrial planets are shown in the table elsewhere in this chapter. The relatively high values of their uncompressed densities suggest that the terrestrial planets are not made of rocks alone, but

	Distance From Sun (10^6 km)	*Equatorial Radius (km)*	*Actual Density (g/cm³)*	*Uncompressed Density (g/cm³, approximate)*
Mercury	58	2440	5.4	5.4
Venus	108	6050	5.3	4.3
Earth	150	6380	5.5	4.2
Mars	228	3400	3.9	3.8

must contain substantial amounts of iron and other high-density elements. Furthermore, there is a general trend of decreasing values with increasing distance from the Sun.

Mercury, the planet closest to the Sun, has the highest uncompressed density and, hence, contains the greatest proportion of iron, in addition to rocks. Venus, Earth, and Mars have progressively lower densities and, hence, contain less iron and more rocks and volatiles.

This is the same pattern observed in the case of the asteroids and comets, and it is yet another indication of the existence of a protoplanetary disk that became differentially heated. Near the Sun the temperatures rose to such high values that only the metals and most refractory minerals survived and became incorporated into the planets. With increasing distance from the Sun the temperatures stayed progressively lower and greater amounts of volatiles survived, giving the more distant planets their lower densities.

The terrestrial planets are not homogeneous bodies of metals, rocks and volatiles. They are differentiated into high-density cores, mainly of iron and nickel, surrounded by rocky mantles of intermediate densities and surface crusts of low-density rocks and volatiles.

These differentiation took place early in the planets formation by mechanisms similar to those that differentiated the larger planetesimals, though the release of energy from gravitational coalescence and from the decay of long-lived radioactive isotopes was much more severe. Figure elsewhere in this chapter depicts the internal structure of the Earth, which is typical (but not identical in details) of all terrestrial planets.

Moon

The Moon is an oddity among the intermediate-sized planetary bodies. Because it is close to Earth and, presumably, was formed

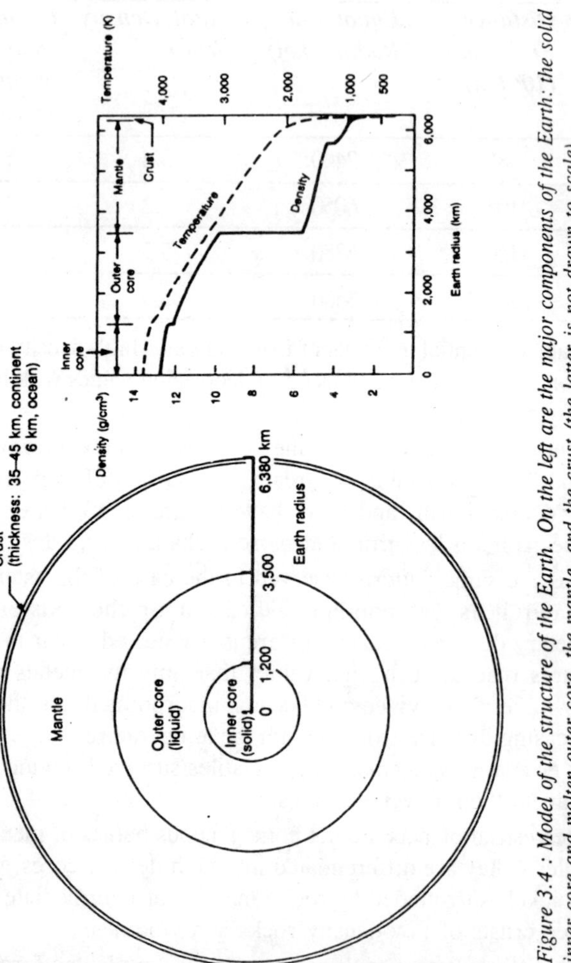

Figure 3.4 : Model of the structure of the Earth. On the left are the major components of the Earth: the solid inner core and molten outer core, the mantle, and the crust (the latter is not drawn to scale).

nearby, its density might be expected to be similar to the uncompressed density of Earth, about 4.2 g/cm³. Instead it is 3.3 g/cm³, lower even than the density of Mars. This indicates that the Moon has either no iron core at all or only a very small one.

There are other anomalies about the Moon's makeup. Its rocks are depleted in the volatiles (H, C, N, O, and the noble gases) relative to Earth. For example, none of the lunar rocks brought back by the Apollo astronauts shows any trace of water, whereas terrestrial igneous rocks (like basalt and granite) generally contain between 0.2 and 0.5% water.

The lunar rocks are also depleted in iron and all elements that tend to combine with iron -the "siderophiles" or "ironloving" elements-such as nickel, cobalt, rubidium, iridium, and platinum.

Clearly, an explanation is required why the Moon has no substantial iron core and why its rocks are depleted in the iron and siderophiles that once must have been part of the material from which it formed. John A. O'Keefe, a NASA astronomer and geophysicist, gives a simple and rather compelling answer (personal communication 1979): "The Moon's iron and siderophiles are in the core of the Earth." He suggests that the material that today comprises the Moon was at one time part of a rapidly spinning Earth.

After the Earth's iron and siderophiles had sunk toward its center to form the terrestrial iron core, the outer layers were torn off by the strong centrifugal forces resulting from the rapid rotation. Others have suggested that the Earth's outer layers were thrown into space by the impact of a very large (perhaps Mars-sized) planetesimal.

However the layers were ejected, part of them -depleted in iron, siderophiles, and volatiles (presumably, many of the volatiles were outgassed and lost due to tidal friction and heating) formed the Moon, and the rest escaped into interplanetary space.

These theories of the origin of the Moon by fission from Earth, originally proposed by Sir George H. Darwin (1845-1912, son of Charles Darwin), or by the impact of a large planetesimal, are not without some difficulties. In particular, they require that either the young Earth rotated unusually rapidly or that it was hit by an uncommonly large planetesimal.

Hence, some planetary scientists prefer the *accretion* theory. According to it, the Moon accreted from material that was in Earth orbit from the outset, rather than part of the Earth. However, this theory cannot explain the Moon's many composition anomalies.

Therefore, in my opinion, further research will eventually prove either the fission or the impact theory to be the correct one. (The theory that the Moon formed somewhere else in the Solar System, perhaps beyond the orbit of Mars, and was captured by the Earth is now generally discounted.)

Galilean Satellites

The densities shown in table elsewhere in this chapter for the Galilean satellites of Jupiter indicate that both Io and Europa have largely rocky structures and, possibly, small iron cores. Photographs show that Io's surface is rocky and covered by sulfur-containing compounds, while Europa is enveloped by a crust of water ice. The low densities of Ganymede and Callisto suggest that they contain substantial amounts of water (perhaps as much as 50% by mass), the rest being rock. They probably have rocky central cores that are overlain by thick mantles of water, either in liquid or solid (ice) forms.

On top of the mantles float hard and rigid crusts of water ice with a thin covering of rock debris from meteorite impacts. The decrease in density in the Galilean satellites with increasing distance from Jupiter follows the same pattern noted for the smaller bodies and the terrestrial planets. It suggests that Jupiter was at one time surrounded by its own protoplanetary-or protosatellitic -disk, that became heated as a result of the planet's formation.

For reasons similar to the ones discussed above, the ensuing temperature gradient was probably responsible for the composition and density differences observed in the Galilean satellites today. Thus, Galileo's suggestion that Jupiter and its four giant satellites are a miniature Solar System was correct beyond the conditions he could envision.

Titan and Triton

The last two planetary bodies to be discussed are Saturn's Titan and Neptune's Triton. The data in table elsewhere in this chapter show that Titan resembles Ganymede in size, distance from its central planet, and density. Hence, astronomers expect that its structure, too, is similar to Ganymede's, though they are'not sure.

But they do know that Titan has a substantial atmosphere, consisting of 80 to 95% molecular nitrogen, with the rest being mainly CH_4 and traces of H_2 and organic molecules. It is believed that these gases are the result of evaporation of ices of H_2O, CH_4, and NH_3 from the satellite's surface, followed by chemical reactions.

Our knowledge of Triton is even more limited than that of Titan, and it will remain so until *Voyager 2* makes its rendezvous with Neptune in 1989. We know that Triton is roughly of the same size as Europa, but its density is smaller. Consequently, ices of H_2O, CH_4, and NH_3 are probably major constituents of this satellite, too. The presence of CH_4 in Triton's thin atmosphere is further evidence of this supposition.

INTERMEDIATE-SIZED PLANETARY BODIES: SURFACE FEATURES

Descriptions of the surface features of the intermediate-sized planetary bodies require selectivity because two decades of space exploration have amassed far more information than can possibly be covered here. Hence, the surfaces of each of the bodies will not be described separately. Instead, they will be compared with each other, paying attention to four topics: (1) impact craters, (2) geological activities, (3) presence of water, and (4) atmospheres. This comparative coverage is consistent with grouping the intermediate-sized planetary bodies into a single family and with the modern trend of regarding the forces that shaped them as part of a continuum.

Impact Craters

The most heavily cratered intermediate-sized planetary bodies are the Moon, Mercury, Callisto, and Ganymede, as well as the southern hemisphere of Mars. Earth, Venus, the northern hemisphere of Mars, and Europa show relatively few impact craters. lo has none at all.

Titan and Triton may be cratered similarly to Ganymede and Callisto, but at present they are not well enough explored for us to be sure. On Callisto, Ganymede, and the highlands of the Moon and Mercury, the crater density is close to the saturation point, with craters edge-to-edge or overlapping and larger craters having smaller ones within them.

Some of these impact craters are truly gigantic as figures elsewhere in this chapter demonstrate. Others are much smaller, and lunar rocks show that they exist even on the microscopic level (see figure 5.16). Very little geological or weathering activity has disturbed these heavily cratered regions, and therefore, the cratering record, much of which dates back nearly 4 billion years, has been preserved..

The Moon and the other heavily cratered bodies are, however, not without some evidence of past geological activity. For example, from about 3.3 to 2.5 billion years ago, lava flows filled in most of the large impact basins on the Moon and formed the dark coloured and relatively smooth "seas" or *mania.* Similar, though somewhat more ancient, lava flows have created the intercrater plains on Mercury.

At the other extreme with regard to crater density lies lo, geologically the most active body in the Solar System. Its volcanic ejectae and lava flows erase craters as rapidly as they form. On Earth, geological activity is somewhat less severe but is still sufficiently dynamic that only a few impact features older than about 500 million years remain.

The same is true of Europa, though on this body it is not the movement of rock that obscures impact craters, but the flow of ice. Mars is interesting. Its heavily cratered southern hemisphere also contains intercrater plains like those on the Moon and Mercury. However, its

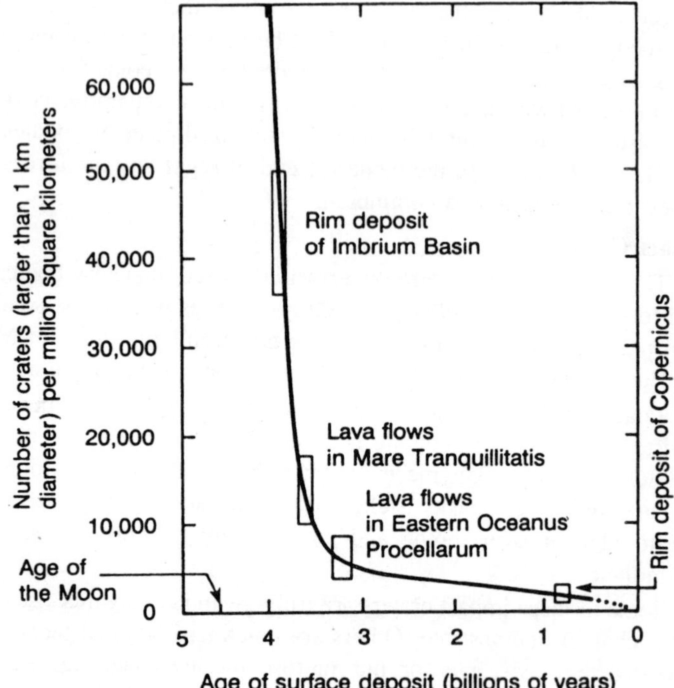

Figure 3.5 : Density of impact craters at four different regions of the lunar surface plotted against the regions ages.

northern hemisphere is dominated by volcanoes and other features of geological activity, which have erased most of the older impact craters.

The collective record of impact craters on the intermediate-sized planetary bodies indicates that up to approximately 3.8 billion years ago all of them were heavily bombarded by large and small planetesimals.

Most likely, this bombardment constituted the terminal phase of the accretion that created these bodies in the first place. From about 3.8 billion years forward, the bombardment declined rapidly; though even today it has not yet reached the zero point, as noted in the discussion of meteorite falls on Earth.

Geological Activity

Geological activity on planetary bodies is caused by surface forces that are sufficiently strong to bring about movements in the crust. The forces may be in the horizontal or vertical directions or both, and they may be pushing, pulling, twisting, contracting, or lifting kinds of forces. They usually are the result of gravity, convective motions inside the body, weathering activities on the surface, and tidal interactions with another nearby body.

The forces are most likely to produce movement if the surface crust is thin and weak, and if the interior is hot and in a molten or plastic state. The interior then acts as a lubricant that enhances movement. In contrast, if the surface crust is thick and strong and if there is little or no melting inside, the surface forces are unlikely to bring about movement. The body is then geologically inert.

The major sources of heat for melting the interior of the intermediate-sized planetary bodies are the gravitational energy that was released during their formation and the energy liberated since then by the decay of long-lived unstable isotopes, such as uranium-235 and -238, thorium-232, and potassium-40.

Both of these heat sources are most effective in the largest bodies, for their surface-to-volume ratios are relatively small (surface μ radius^2, volume μ radius^3; hence, surface-to-volume ratio μ 1/radius), so that they lose their internal heat slowly. Also, much more heat was released early in the bodies' histories than, for example, today because the stored-up heat from their formation and the amount of unstable isotopes were greatest then.

Io

lo is one of the few bodies in the Solar System whose geological activity is mainly due to tidal heating. This heating is brought about by gravitational interactions with nearby Jupiter, which creates an appreciable tidal bulge on lo.

As lo revolves about Jupiter and alternately approaches and recedes from it due to a slight eccentricity in its orbit (the eccentricity is maintained by gravitational interaction with Europa), the tides raise and lower different parts of lo. This causes friction by rock rubbing against rock and produces enough heat to melt part of Io's interior.

Unbalanced forces, probably also due to the tides, open up cracks in Io's crust and allow the molten materials of the interior, consisting largely of sulfur and sulfur-containing compounds, to push to the surface. There they gush in geyserlike fashion into the open, creating enormous lava flows and huge volcanoes, giving the satellite its mottled appearance. Unquestionably, lo deserves its reputation as the Solar System's strangest and geologically most active planetary body.

Earth

After lo, the geologically most active intermediate-sized planetary body is Earth. This is not surprising. It is the most massive body among them. Thus it has the greatest amount of stored-up internal heat, is molten to a greater extent than the others (except for Io, because of its tidal heating, and, possibly, Venus), and its outer solid layer, the lithosphere, is relatively thin.

In fact, the terrestrial lithosphere is a mere 100 km thick on the average, compared to a total radius of 6380 km. The terrestrial lithosphere is composed of a number of distinct and rigid tectonic plates that float like rafts on the denser underlying asthenosphere. Forces within the Earth push and pull on the plates, slowly moving them about relative to each other.

This pushing and pulling is called *plate tectonics* and is responsible for most volcanism and earthquakes, as well as for the building of mountains and the shaping of continents and ocean basins.

In addition to plate tectonics, the surface of the Earth is continuously altered by weathering. Flowing water, glaciers, and wind erode the land, and rivers carry the resulting debris toward the oceans, where they deposit it as sediments. Over time, the layers of sediments become compressed and solidified as more sediments are piled on top of them.

When subsequent uplifting and erosion expose these ancient deposits, their banded structures reveal their mode of origin. Examples are the banded structures of the Canadian Rocky Mountains and the Grand Canyon.

Plate tectonics and weathering have been going on on Earth for at least 4 billion years. The oldest surviving rock formations of these processes are the *continental shields,* which are present on every continent and comprise about 10% of the Earth's surface. These shields are the closest terrestrial analogues to the cratered highlands of the Moon, Mercury, and the southern hemisphere of Mars, though few craters remain.

Venus

Venus has nearly the same mass and density as Earth. Hence, we might expect that plate tectonics and volcanism were, and, perhaps, still are, also important in sculpturing its surface. Unfortunately, we do not yet know for certain if this is the case because the planet's surface remains hidden under a thick, opaque atmosphere.

However, radar soundings bounced off Venus from Earth stations and orbiting spacecraft reveal a topography not unlike that of Earth and hence point to geological activity. There are two massive highlands, Ishtar Terra and Aphrodite Terra, that resemble terrestrial continents.

There also are several smaller islandlike regions and broad depressions comparable to the ocean basins on Earth.

Most of these landforms are shallower than the contours on Earth, but not all. For example, Ishtar Terra contains a number of substantial mountain ranges. The highest of them, Maxwell Mons, rises 12,000 m above the planet's average radius and is comparable to the Himalayan Mountains.

Mars

Like Earth and Venus, Mars has had a complex and interesting geological history. Its southern hemisphere is heavily cratered and shows evidence of interspersed flooding by lava. The northern hemisphere also contains craters and intercrater lava plains, but to a much smaller extent.

Its most conspicuous features are enormous mountain ranges and canyon systems that dwarf their counterparts on Earth. The largest of the mountain ranges is a 6000-km-long uplift called the Tharsis Ridge that is punctuated by four gigantic volcanoes. The volcanoes bear a striking resemblance to the shield volcanoes on Earth, such as Mauna Kea and Mauna Loa of Hawaii.

The largest of them, Olympus Mons, measures 600 km across at its base and rises 26,000 m above the surrounding terrain. To the east of the Tharsis Ridge lies a 4000-km-long system of interconnected canyons called Valles Marineris. Some individual canyons are more than 200 km wide and 7000 m deep.

They show modifications by landslides, wind erosion, and flowing water, mud, or, possibly, glaciers (though no liquid water exists on the Martian surface today; see below). In comparison, Earth's Grand Canyon is 450 km long, 30 km across at its widest point, and 2000 m deep.

The lava flows, mountains, and canyons of Mars indicate that, with regard to geological activity, this planet lies somewhere between the Earth and the Moon. After having been heavily cratered early in its history, its geology began to evolve like that of the Earth.

However, before the entire planet was affected, its geological activity came to a halt. Most likely, the reasons were the relatively small mass and size of Mars, which, respectively, are one-tenth and one-half those of Earth.

Consequently, Mars was heated less initially; its crust became thicker than that of Earth; and its volcanism and tectonic activities stopped early in its history.

Europa

The only intermediate-sized planetary body besides Earth, Venus, and Mars that is suspected to have had recent and, maybe, current tectonic activity is Europa. The basis for this speculation is its lack of any large (over 100 km) and, hence, old craters.

Furthermore, Europa's surface is marked by a vast tangle of light and dark streaks that are probably filled-in cracks and indicate the presence of tensional forces pushing and pulling on the crust. If tectonism does exist on Europa, the tectonic plates would not be rock, like those on Earth, but layers of water ice.

The heat necessary for keeping the underlying rocky layers molten so that tectonic forces can produce crustal movement would largely come from tidal interactions with Jupiter, though to a much smaller degree than on lo.

Mercury, Moon, Ganymede, Callisto

The remaining intermediate-sized planetary bodies -Mercury, Earth's Moon, Ganymede, and Callisto -are *one plate planets* (too little is known of Titan and Triton to say anything about their geological activities). No large-scale tectonism deforms their surfaces today, nor is there much evidence of volcanism.

However, the presence of ancient intercrater lava plains on Mercury and the Moon does indicate that during the first half of their existence molten magma managed to penetrate from their interiors to the surface. Likewise, the rather complex system of grooves and ridges that crisscross Ganymede, with estimated ages of about 3.5 billion years, suggests that this body, too, has experienced some form of geological activity. It appears that by the time the Solar System was 2.5 billion years old, the crusts of all of these bodies began to harden into single, unyielding plates and their geological histories drew to a close.

Presence of Water

Water is present today on all intermediate-sized planetary bodies with the exception of Mercury, the Moon, and lo.

Mercury, Moon, Io

Mercury and lo have no water because of their proximities to the Sun and Jupiter, respectively, and the heating they experienced during their formation. The Moon is without water probably because of the way it formed.

Venus

Venus is also very dry. The little water it does possess exists as vapor in its atmosphere. However, at one time it may have had as much water as Earth, but lost most of it as a result of the evolution of its atmosphere. The evidence comes from the abnormally high ratio of deuterium to hydrogen (^2H:^1H) in the water that is still present.

This discovery was made in 1978 with an instrumented capsule that descended through the Venusian atmosphere, after it was delivered to the planet by the United States spacecraft *Pioneer Venus Multiprobe*. The high ^2H:^1H ratio suggests that early in Venus's history UV radiation from the Sun broke the planet's water molecules into their atomic constituents, ^1H, ^{16}O, and an occasional ^2H.

The oxygen reacted with other compounds and is now locked up in the crust of Venus. The heavy isotope of hydrogen, ^2H, also was retained, but most of the ^1H escaped into space because of its small mass. This raised the ^2H:^1H ratio in the remaining water to roughly 100 times its value on Earth.

Europa, Ganymede, Callisto, Titan, Triton

Of the intermediate-sized planetary bodies that have retained appreciable amounts of water, the greatest amount is found in the outermost ones-Europa, Ganymede, Callisto, Titan, and Triton -because of the low temperatures that prevailed during and after their formation. In fact, water may constitute up to 50% of their masses, excepting only Europa.

Differentiation early in the histories of these bodies forced the water out of their cores to form massive mantles surrounded by thin crusts. The surfaces are now at temperatures of 150 K or less and are frozen into hard, rigid ice that is capable of retaining the shapes of impact craters and other surface features for considerable lengths of time.

For example, Ganymede and Callisto are heavily cratered, much like the rocky crusts of Mercury and the Moon, though the ice craters are considerably shallower (because ice slowly flows and deforms, as glaciers on Earth demonstrate).

Earth

A little more than two-thirds of the surface of Earth is covered by oceans. From this we might conclude that, concerning the abundance of water, Earth falls somewhere in the middle between those bodies that consist half of water and those that have none. That is a serious misconception. The water on the surface of the Earth adds up to approximately 2.3×10^{-2} % of the total mass of our planet. Several times this amount of water may be locked up in the rocks of the crust and mantle (basalt and granite contain between 0.2 and 0.5% water). But Jthe total can hardly be more than a fraction of a percent of the total mass of the Earth.

Hence, with regard to water, Earth resembles barren Mercury much more than water-laden Ganymede or Callisto. Clearly, the planetesimals from which Earth was assembled had lost most of their water or never possessed much in the first place. Otherwise Earth should have ended up with roughly the same proportion of water as the outer planets.

Because the water loss was so great, we may assume that any number of minor changes in the protoplanetary disk -such as slightly more or less heating by the nascent Sun-could have significantly altered the amount of water that did end up on Earth, with important consequences for the environment and the origin and evolution of life.

Mars

Though Mars is more distant from the Sun than Earth, and hence should be much wetter, the only water apparent on its surface today is in thin layers of ice that cover its poles. These icecaps grow and wane with the seasons, subliming (that is, changing from the solid to the gaseous state without passing through the liquid state) during the summer and recondensing during the winter.

If liquid water were present on the Martian surface, it would not be stable, given Mars's present temperature range of 150-290 K and a barometric pressure of a mere 7/1000 that of the sea level pressure on Earth. It would either boil or freeze, depending on the temperature.

Still, there is ample evidence that at one time liquid water flowed on Mars. For example, the cratered terrain is cut by countless *runoff channels* that bear some resemblance to terrestrial valleys eroded by flowing water. These channels start out small, have numerous tributaries, increase in size downstream, and are, on the average, a few tens of kilometers long.

Most were probably cut between 3.0 and 3.5 billion years ago, shortly after the heavy initial cratering ended. Other kinds of channels, the *outflow channels,* have the markings of sudden flooding by huge quantities of water or other kinds of fluvial matter. These channels are much bigger than the runoff channels, frequently measuring tens of kilometers across, and they meander for hundreds of kilometers across the Martian terrain. They usually are flatbottomed, often contain islands, and generally start abruptly and full-grown.

It is as if huge reservoirs of water had suddenly burst their dams and emptied. In fact, there is some evidence for the previous existence of large standing bodies of water in some regions, in the form of sedimentary layering, but their relation to the outflow channels is ambiguous. Underground reservoirs of water, soil saturated with water that suddenly became unstable, and glaciers have also been suggested as likely sources of the flooding.

The outflow channels are younger than the runoff channels and probably were formed when the geological activity on Mars reached its peak, roughly 2.5 billion years ago. The geological activity was accompanied by outgassing and the release of large quantities of water and other volatiles, as occurred on Earth. Much of the water was probably lost to space because of the low surface gravity of Mars. A small fraction is still present in the planet's polar icecaps. The rest is suspected to exist as permafrost or possibly in liquid form under the Martian surface.

Atmospheres

Of the intermediate-sized planetary bodies, only Venus, Titan, and Earth have substantial atmospheres, followed by Mars and Triton. The major gases in the atmospheres, the surface pressures, average surface temperatures (in degrees Kelvin and Celsius), and surface gravities of these bodies are listed in table 5.3. In comparison, the other intermediate-sized planetary bodies have barren rocky or icy surfaces.

The sizes of planetary atmospheres are determined by two factors: (1) the sources that bring gases to the surface, and (2) the loss mechanisms that tend to remove them. The sources of atmospheric gases are tectonism and volcanism. They deliver severely heated, molten rock to the surface and expose them to low pressures, which allows volatiles trapped in the rocks to escape or outgas into the open.

For example, H_2O, CO_2, and N_2 are the most common volatiles trapped in the surface and upper mantle rocks of the Earth, as the compositions of the clouds billowing forth from terrestrial volcanoes demonstrate. Hence, H_2O, CO_2, and N_2 were the major original constituents of our planet's atmosphere.

A second source of atmospheric gases is evaporation of liquids or sublimation of ices from the planetary surfaces. The mechanisms by which planets lose atmospheric gases are (1) diffusion into interplanetary space if their surface gravities are small or the atmospheres are very hot, and (2) loss to the ground. The loss to the ground may occur either by condensation, such as the falling of rain, or by chemical reactions that bind the gases to the surface rocks. The various gain and loss mechanisms of planetary atmospheres are interdependent and far too complex to be discussed here in detail. Briefly, Earth and Venus, for example, have substantial atmospheres because of the extensive geological activities they experienced throughout most of their pasts. Furthermore, both planets are massive and have relatively strong surface

gravities with which they retain their atmospheres. On Mars, by contrast, the early geographical activity was interrupted and, thereby, so was the growth of its atmosphere. Besides, Mars is less massive than Earth or Venus, with only about one-third their surface gravity. Hence, atmospheric gases escape to space more easily.

Mercury is too hot and the other terrestrial planetary bodies have surface gravities that are too weak to have accumulated appreciable atmospheres. The one exception is Titan. Its surface gravity and temperature are comparable to those of Ganymede and Callisto, but unlike these two bodies it has a very thick atmosphere.

In fact, above a given area of its surface, Titan's atmosphere exceeds that of Earth by a factor of ten. At present we do not know why this is so.

THE JOVIAN PLANETS

The Jovian planets-Jupiter, Saturn, Uranus, and Neptune -are the "gas giants" of the Sun's planetary system. They possess no solid surfaces and no ancient impact craters or deformations by geological activities mar their surfaces. Instead, thick atmospheres envelope them, whose top layers are characterised by gigantic and often colourful wind and storm patterns.

The bottom layers of the atmospheres pass smoothly over into liquid mantles, and near their centers they possess cores of mainly rocks and iron. Thus, the Jovian planets are quite unlike the intermediate-sized planetary bodies and deserve to be grouped into a family of their own.

Their unique structures stem from their large masses and sizes in comparison to the other planetary bodies and from the volatiles they possess in great abundance. For example, Jupiter has roughly 300 times the mass and 10 times the size of the Earth, and Saturn is not much smaller, as table 4.1 indicates.

Furthermore, the compositions of Jupiter and Saturn are dominated by hydrogen and helium, followed by carbon, nitrogen, and oxygen. Uranus and Neptune are less massive and smaller than Jupiter and Saturn, but they, too, consist largely of volatiles, though they lost much of their original hydrogen and helium. The large abundances of volatiles in these four planets follow from both their observed surface compositions and from their unusually low average densities.

The combined mass of the Jovian planets adds up to 0.13% of the mass of the Sun and it exceeds the mass of the other planets, satellites, and smaller bodies of the Solar System by about a factor of 200. Thus, next to the Sun, the four Jovian planets are the Solar System's major components.

Core and Mantle Structure

The three-layered structures of the Jovian planets -gaseous atmospheres, liquid mantles, and solid or liquid cores -differ considerably from one planet to the next because of differences in their masses, compositions, and distances from the Sun.

The smallest difference in structure among the Jovian planets is found in the makeup of their cores. Model calculations indicate that the cores of all four planets consist mainly of silicates and iron, with small admixtures of other heavy elements. Clearly, some form of composition differentiation, similar to the one in the intermediate-sized planetary bodies, took place in these giant planets.

Uranus and Neptune probably have the largest cores in proportion to their total sizes because they have lost the most volatiles. In contrast, Jupiter and Saturn have the highest core pressures and temperatures because they are the most massive. The estimated conditions at their centers, compared to the Sun's, are as follows:

	Jupiter	*Saturn*	*Sun*
Temperature (K)	25,000	20,000	16×10^6
Pressure (earth atmospheres)	8×10^7	5×10^7	3×10^{11}
Density (g/cm^3)	20	15	160

The mantles of the four planets are liquid, but beyond that they show a great deal of variation. The mantles of Jupiter and Saturn consist mainly of hydrogen and helium and have two-layered structures. In the inner layers, high pressures keep the hydrogen molecules stripped of some of their electrons, which produces metallic characteristics, such as high conductivity of heat and electricity.

Hence, these layers are commonly referred to as the *metallic layers*. In the outer mantle layers, which are at lower pressures, the molecules are not stripped of electrons and, consequently, they are not metallic -they are just ordinary liquids. The mantles of Uranus and

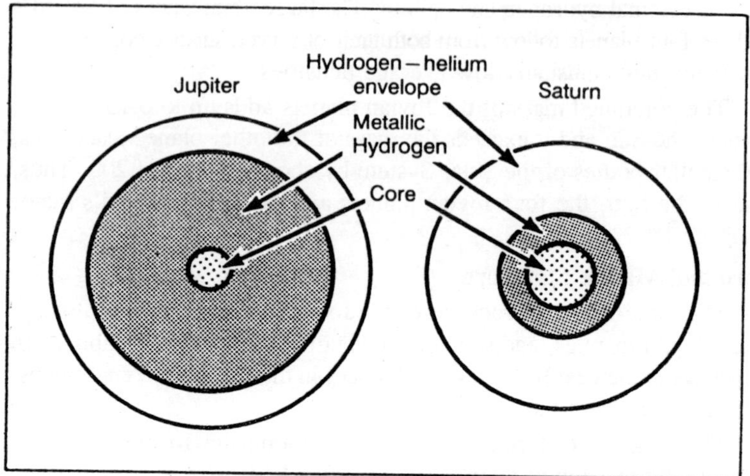

Figure 3.6 : Approximate models of the interior structures of Jupiter and Saturn.

Neptune are believed to be single-layered and to consist chiefly of liquid water, methane, and ammonia, without any metallic characteristics.

Atmospheres

The greatest differences in structure among the four planets arise in their atmospheres. Jupiter's and Saturn's atmospheres are characterised by welldefined coloured zones or belts that encircle the planets and rotate with them. These zones are remarkably stable and have not changed their latitudinal positions during the last 80 years.

Close-up photographs sent back by the *Voyager I* and *Voyager 2* spacecrafts show that the zones of both planets are in a general state of turmoil, with large and small eddies, fluctuating atmospheric motions, and elongated fast-moving gas streams abounding. The most prominent of these dynamic features have the appearance of dark or bright spots and are useful markers for tracking the planets' global circulation patterns.

The atmospheric compositions are H_2 and He, roughly in solar proportions, followed by small admixtures of NH_3, CH_4, and, in the case of Jupiter, H_2O and various kinds of simple organic compounds such as C_2H_6 (ethane), C_2H_2 (acetylene), and HCN (hydrogen cyanide).

The atmospheric zones of both Jupiter and Saturn are closely associated with alternating westerly and easterly winds. For example, if in one zone the atmospheric winds blow toward the east, those in the two adjacent zones blow toward the west.

These winds are analogues of the jet streams and trade winds on

Earth, which also flow in adjacent zones and in opposite directions. Earth has only two such wind zones (jet streams and trade winds) in each hemisphere, but Jupiter and Saturn have several and they are much wider. For example, the winds of their equatorial zones are approximately 30,000 km and 80,000 km wide and they blow eastward with speeds of up to 500 km/hr and 1800 km/hr, respectively. In the case of Saturn, this is two-thirds the local speed of sound.

The bright and dark spots on the surfaces of Jupiter and Saturn are eddies of circulating atmospheric gases, comparable to the rotating high- and lowpressure systems that determine the weather on Earth. The smaller eddies do not last long.

Within days of their origin they usually are caught between oppositely flowing zonal wind currents, and the resulting shear forces tear them apart. Sometimes they are absorbed by clouds and may be disgorged hours or days later. Complex patterns of oscillating vortex motions are often set up in the wakes of these atmospheric disturbances. Eventually, the eddies and their fragments disappear altogether, only to be followed by new ones that suddenly appear.

The larger eddies last much longer. The most famous of them, the Great Red Spot in the southern hemisphere of Jupiter, has been observed since the invention of the telescope more than 300 years ago and, probably, has existed considerably longer than that. The three white ovals just to the south of the Great Red Spot were first observed in 1938; other large eddies have lasted for months or years.

They are so stable because instead of being sheared apart, they roll along between the wind currents of adjacent zones. In contrast to Jupiter and Saturn, very little is known about the atmospheric conditions of Uranus and Neptune. These two planets are too distant and faint to reveal much of their constitution to earth-bound observations.

However, in January, 1986, *Voyager 2* passed Uranus and it is now on its way to Neptune, which it will reach in August of 1989. Uranus is odd in that its equatorial plane as well as its ring system and the orbits of its 15 satellites are nearly perpendicular to the plane of the ecliptic.

In contrast, the equatorial planes of all other planets are aligned much more closely with the ecliptic. In fact, at present Uranus's south pole points almost directly toward the Sun. Data sent back by *Voyager 2* indicate that Uranus rotates once every 17.3 hours.

Its atmosphere contains hydrogen and helium in proportions similar to those of Jupiter and Saturn. It also contains small admixtures of methane and various other hydrocarbons. There is evidence of cloud structure and atmospheric winds, but *Voyager 2* detected no bands or other detailed surface characteristics.

The planet is shrouded in a permanent smoglike haze, which hides most of the atmospheric features that lie below. The Uranian temperature is 50 K at a level where the pressure equals 0.1 earth atmospheres, and it increases with depth. Surprisingly, the temperature varies very little between the planet's sunlit and dark sides.

This is further evidence of the existence of atmospheric circulations, which distribute the heat received from the Sun. Neptune, the most distant and faintest of the Jovian planets, shows up as a greenish disk when viewed through earth-bound telescopes. Its known atmospheric constituents are hydrogen, helium, and methane. Virtually no surface markings are discernible from Earth. Few other details are presently known about this planet.

ORIGIN OF THE SOLAR SYSTEM

Now that we have discussed the orbital distribution and physical structures of the major components of the Solar System, let us turn to their origin. This is a topic of age-long interest and was one of the first to be examined during the early development of western science, more than 300 years ago. In 1644 Rene Descartes, the great French philosopher, suggested that at one time the Sun was surrounded by a large rotating disk of gas from which the planets and their satellites formed.

One hundred years later, this *nebular theory* was challenged by the French naturalist Georges Louis Leclerc de Buffon (1707-1788). He proposed *a collision theory* according to which a massive comet or, as proposed later, a star passed close to the Sun and ripped out the material from which the planets and their satellites then assembled.

Neither of these two theories was based on sufficient facts or theoretical reasoning to be fully convincing, and during the succeeding centuries first one and then the other gained favour. The German philosopher Immanuel Kant (1724-1804) and the French mathematician Pierre Simon Laplace (1749-1827) supported the nebular theory.

Laplace pointed out that a rotating disk of gas would slowly contract and develop rings of materials from which planetary bodies could form. Throughout much of the nineteenth century, this remained the accepted theory among most scientists.

However, in the second half of that century, the British physicist

James Clerk Maxwell (1831-1879) calculated that shear forces resulting from the differential rotation of the disk would prevent the matter from condensing into individual bodies.

Thus, Buffon's collision theory was resurrected, and during the first few decades of the present century the English physicist Sir James H. Jeans (1877-1946) and others developed it further. Because a collision between two stars is an extremely rare event, they argued that our planetary system must be unique or at most one among a few.

Maxwell's critique of the nebular theory was based on the mistaken notion that stars and the interstellar material have the same composition as the Earth. Only in the late 1930s was it discovered that in fact hydrogen and helium are the major constituents of the Cosmos and that material like that of rocks adds up to no more than about 2%.

From this new knowledge, the German physicist Carl Friedrich von Weizsacker concluded that the primitive circumsolar disk must have been much more massive than are the planets today, and he demonstrated that Jeans's objections to the nebular theory were inapplicable. It was also realised that material ripped out of the Sun by a passing star would disperse or fall back onto the Sun, rather than condense into planetary bodies.

These new theoretical insights led to the abandonment of the collision theory and the general acceptance of the nebular theory. Since the 1940s, an increasing number of researchers have worked on the nebular theory of the origin of the Solar System, using the growing knowledge about interstellar clouds and star formation, building numerical models with the aid of high-speed computers, and, more recently, applying the wealth of new observational data gathered from space.

The theory is still far from complete, with many rival ideas about some of the details. However, there is a general consensus today among the experts how, in broad terms, the Sun and its planetary system formed from an interstellar cloud of gas and dust.

Stage I

About 4.5 to 4.6 billion years ago, along one of our galaxy's spiral arms, a supernova exploded near an interstellar cloud of gas and dust. As its shock front swept through the cloud, it compressed many of its dense subregions to the point of gravitational collapse.

One of the dense clumps thus formed contained the raw materials—hydrogen (73.6% by mass), helium (24.8%), and heavier elements (1.6%)-for the birth of the Solar System. Before compression, the clump was

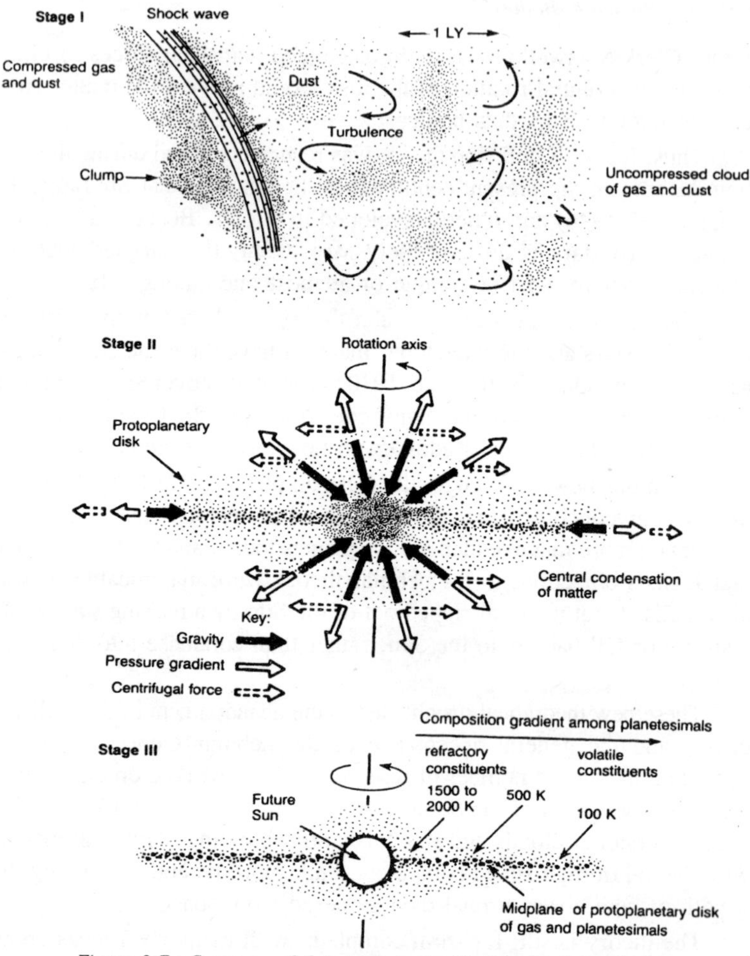

Figure 3.7 : Summary of the stages in the birth of the Solar System.

roughly 0.1 to 1.0 LY (approximately 10^3 to 10^5 AL) across and it had a mass between 1.5 and 2 solar masses. After compression, its size was reduced by roughly a factor of ten, though there is at present little certainty about these values.

The clump rotated slowly; its temperature was about 5-10 K; and most of its material was condensed into molecules and dust grains. Some of the molecules and dust grains had arrived with the shock wave from the supernova and differed in composition from the bulk material of the clump.

For the sake of balance, be aware that some scientists are not

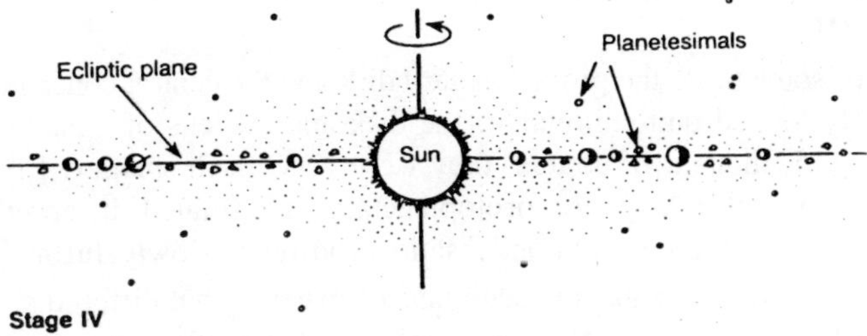

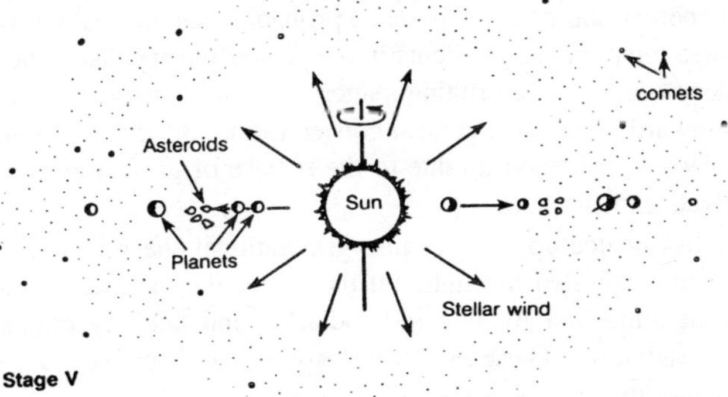

(Figure 3.7 contd.)

convinced that a supernova was responsible for the compression of the cloud to the point of gravitational collapse. They think that this point was reached by the gradual and natural evolution of the cloud, without any external, compressive forces. However, their theory does not readily account for the range of isotopic abundances found in the Solar System today.

Stage II

As gravity pulled the clump together, it began to rotate faster. Three kinds of forces acted on the clump: gravity, the pressure gradient, and the centrifugal force resulting from rotation. (A fourth possible force—the force from magnetic fields—is disregarded here.)

Gravity pulled the matter radially toward the clump's center, the pressure gradient pushed it outward, and the centrifugal force pushed it in directions perpendicularly away from the clump's rotation axis, as shown in figure elsewhere in this chapter. The net result was that the clump slowly contracted into a disk—*the protoplanetary* disk-revolving around a growing central concentration, the future Sun.

Stage III

For some time, the protoplanetary disk and the central concentration probably looked quite similar to the disk and nucleus of galaxies like NGC 3115, though, of course, they were very much smaller. Like the disk of a spiral galaxy, the protoplanetary disk rotated differentially, faster close to the central condensation and more slowly further out.

As a result, neighboring segments of material that differed slightly in distance from the rotation axis orbited at slightly different speeds and, therefore, rubbed against each other. This created friction (that is, shear forces) and led to local turbulence.

Yet another source of turbulence probably was thermal convection due to steep temperature gradients in the protoplanetary disk. The effect of turbulence was that the orbiting matter was slowed down and gradually spiraled inward toward the central concentration. As that concentration of matter grew, it heated up due to the release of gravitational energy and evolved into the Sun.

The disk heated up also because gravitational energy was released in it as matter spiraled through. Furthermore, the disk was heated by the nascent Sun, first by infrared radiation and later by optical and ultraviolet radiation. These heat sources were far stronger close to the Sun than farther out.

Consequently, a temperature gradient was established from about 1500-2000 K in the region where Mercury eventually formed to approximately 100 K or less in the disk's outer fringes.

In the course of the growth of the protoplanetary disk, the concentration of its matter increased enormously. This was particularly the case along the disk's midplane, where a number of important and irreversible changes began to take place.

The orbiting dust particles were so densely packed there that they collided very frequently with each other and coalesced into growing bodies, first of pebble size and then of boulder size and larger. Quite rapidly, many of the bodies reached diameters of tens to hundreds of kilometers and became the planetesimals, or "little planets."

At first, when the planetesimals were still very small, they had a rather loose and fluffy structure, similar to that of the dust grains. However, as they grew in size, self-gravity compacted them into hard, solid objects. Simultaneously, heat released by gravitational contraction and the decay of unstable isotopes, such as ^{26}Al, caused melting and differentiation.

The growth of the planetesimals was strongly affected by the disk's temperature. In the inner parts of the disk, where temperatures reached 1500-2000 K, the coalescing dust particles lost all but their most refractory constituents. Consequently, the planetesimals that formed there consisted only of silicate rocks and metals, without any volatiles.

Toward the middle regions of the disk, corresponding to where Mars and the asteroids eventually formed, the temperature rose to only about 500 K. Hence, the dust particles retained many of their volatiles (such as H_2O, CO, CO_2, H_2S [hydrogen sulfide], N_2 and some organic molecules) and carried them along in the assembly of the planetesimals.

Finally, in the outer regions of the disk that were heated the least, virtually all volatiles stayed with the dust particles, including those most easily vaporised, such as CH_4 and NH_3, and became part of the planetesimals. Thus, the temperature gradient of the disk led to a composition gradient among the planetesimals, from those with only refractory constituents to those composed of rocks intermingled with progressively more volatiles at increasing distances from the nascent Sun.

Stage IV

When the growth of the planetesimals first began and for some time thereafter, the gaseous component of the disk (consisting mostly of hydrogen and helium) was still thick and dense, and gas continued to spiral through it toward the Sun. The planetesimals participated only very little in this spiraling motion because they were massive and compact.

They plowed right through the gas without being affected much and traveled in nearly circular or elliptical orbits around the Sun. Most of their orbits lay close to the midplane -the future ecliptic plane of the protoplanetary disk, but some were inclined to various degrees. Large eccentricities and inclinations in the orbits were the result of continuous and ever-varying gravitational pulls that the planetesimals exerted on each other as they orbited the Sun.

Planetesimals that were in elliptical and inclined orbits intersected the orbits of other planetesimals and, sooner or later, had near encounters or collisions with them. Some of the collisions were sufficiently energetic to shatter the planetesimals and disperse part or all of the resulting debris.

The heavily cratered surfaces of many of the small and intermediate-sized bodies in the Solar System today are evidence of this violence. However, most often the collisions were rather gentle because nearcircular and little-inclined orbits predominated. Hence, most collisions led to merger and growth. The planetesimals that grew most rapidly were those that by chance acquired above-average masses early on and thus were most successful in gravitationally attracting and assimilating other smaller planetesimals.

As illustrated by a sequence of computer simulations made by George W Wetherill of the Carnegie Institution of Washington, dominant planetesimals established themselves at regular orbital intervals. In the course of time, they swept up nearly all of the other smaller bodies in their orbital neighborhoods and evolved into the terrestrial planets we observe today.

Collisions and accretion of solid planetesimals, as described so far, were the principal mechanisms of planet formation only in the inner part of the Solar System. In the outer parts, where temperatures were lower, planet formation was greatly influenced by the gas component of the protoplanetary disk.

In fact, hydrogen and helium became the major constituents of Jupiter and Saturn and, to a smaller degree, became incorporated in Uranus and Neptune as well. At present it is not known whether, in the case of these four planets, solid planetary cores formed first, followed by the gravitational acquisition of envelopes of hydrogen and helium; or gaseous spheres formed first, followed by differentiation and the settling of rocky components toward their centers.

Whichever sequence was the actual one, all four giant planets probably were, early in their formations, surrounded by their own protosatellitic disks of gas and dust, with temperature and composition gradients, from which satellites formed.

Stage V

While the planetesimals and planets were forming, the young Sun was also evolving. Within several hundred thousand to a million years of the initial compression of the primordial clump, the Sun became a T Tauri star and experienced the surface eruptions, rapid variations in light output, and mass loss in the form of a strong wind that are typical of that evolutionary stage.

The wind blew for millions of years and swept away all of the gas and dust that had not yet coalesced into planets, satellites, or planetesimals, exposing these bodies to interstellar space.

Orbits, Rotations, and Composition Gradients

Some of the specific features of the Solar System can be explained in terms of the formation theory just described, but some cannot.

Because the planets were the result of accretion of many small bodies and, in the case of the Jovian planets, of initially widely distributed gaseous material, their final orbits turned out to be averages of the individual orbits of the many small bodies and the gas.

These averages added up to the near-circular and little-inclined orbits we observe today. The planets directions of revolution around the Sun were also the same as the rotational direction of the protoplanetary disk, namely from west to east. The only exceptions were Mercury and Pluto, both of which acquired above-average orbital inclinations and eccentricities.

We do not know what caused these results. However, the case of Pluto is not surprising, for it is very small and resembles more an asteroid than a genuine planet.

Most of the collisions between planetesimals were probably not head-on, but off-center or at glancing angles. The same was most likely true of the coalescence of gases in the formation of the outer planets. Thus, all planets ended up with net rotations that, with the exceptions of Venus and Uranus, were also in the direction from west to east.

However, it is not clear at present why the planets have the particular rotation rates we observe, or why two rotate from east to west and the rest from west to east.

The development of a composition gradient among the planetesimals combined with the substantial amounts of hydrogen and helium the outer planets managed to acquire, explains the different compositions in the planets today.

The Earth, for example, formed in a part of the protoplanetary disk in which heating was so severe that most of the volatiles that originally were present in the dust grains were vaporised and driven off. Consequently, the Earth acquired very few volatiles, compared to its bulk constituents of silicate rocks and iron-nickel metals.

Most of the volatiles it possesses today-such as those constituting its ocean and atmosphere, and the CO_2 that became buried as dolomite and limestone in its crust -were probably brought to it by planetesimals from the vicinities of Mars and the Asteroid Belt.

Perturbations of the planetesimals' orbits by near encounters with other planetesimals and nearby Jupiter deflected them inward toward Earth, which then captured them.

As Wetherill's computer diagrams show, this kind of scrambling of the planetesimals' orbits was common in the protoplanetary disk. Thus, each planet had a much wider "feeding zone," from which it drew its raw materials, than today's spacing between planetary orbits. The planet that had the greatest effect on the scrambling and broadening of planetesimal orbits was Jupiter with its large mass. It perturbed the planetesimals in its immediate vicinity so strongly that, for example, at 2.8 AU from the Sun no planetary mass accumulated where, according to the Bode-Titius law, a planet should revolve.

Most of the planetesimals of that region were dispersed to other parts of the protoplanetary disk. The few that survived there are today's asteroids. Likewise, Jupiter's perturbing effect removed many of the planetesimals from the neighborhood of the future planet Mars (at 1.5 AU), which explains why this planet acquired only about one-tenth the mass of Earth.

Furthermore, many of the comets in the Oort Cloud, which extends 100,000 AU outward from the Sun, are believed to have been deflected there by Jupiter and the other giant planets.

The deflection of planetesimals from one part of the Solar System to another was responsible for the heavy bombardment that the planets experienced for hundreds of millions of years after they were formed. Only as the supply of planetesimals ran low, about 3.8 billion years ago, did the bombardment diminish. However, even today, the supply - the asteroids and comets—is not fully gone and the bombardment continues, though at a very low rate.

Despite the many features of the Solar System that are explained, at least qualitatively, by the theory just discussed, there remain numerous important and unresolved questions. For instance, how did it happen that a few, relatively tiny, planets swept up nearly all of the material between them? Or, put differently, why are there not more planets, each with a smaller mass? Why are the masses of Jupiter and Saturn roughly one order of magnitude greater than those of Uranus and Neptune, and two orders of magnitude greater than those of Earth and Venus?

What was the time scale of planet formation? Was the main assembly phase completed in a hundred million years? Or did it take much more or much less time? Are planetary systems inevitable by-products of star formations? Are they found only around single stars, or can binary stars have planets, too (perhaps in large orbits around close binaries and in relatively small orbits around widely spaced binaries)? Only by continuing investigations of the Solar and other stellar systems-both observationally and theoretically (in particular, through computer modeling)-can answers be found to such questions.

The origin of the Solar System dates back some 4.5 billion years, when the protosolar nebula contracted under the influence of its own gravitational forces. The greatest condensation of matter occurred at the center of the nebula and gradually evolved into the Sun.

Orbiting the central condensation was a disk of gas, dust, and rapidly growing planetesimals. Collisions and coalescence among the planetesimals led to the formation of the small and intermediate-sized planetary bodies, while coalescence of the gas (at distances of Jupiter's orbit and beyond) produced the Jovian planets.

All remaining uncondensed gas and dust were driven off into interstellar space by the strong wind that blew from the young Sun as it evolved through its T Tauri stage. After the bodies of the Solar System were formed, they continued to evolve. In the case of most of them this evolution was strictly physical. For example, the Sun kept burning hydrogen and pouring fourth radiant energy. Venus, Earth, Mars, and most other bodies of intermediate size experienced major geologic changes and some acquired atmospheres.

Many of the asteroids collided, some of them with the planets. The member bodies of the Oort Cloud, in the outer fringes of the Solar System, were periodically perturbed by gravitational forces of uncertain origin, and some of them were sent, as comets, in elongated orbits toward the inner part of the system.

There, after numerous circuits around the Sun, they either crashed on one of the planets or the Sun, or they broke up and became part of the interplanetary debris. However, the most remarkable evolution in the Solar System was not physical; it was biological. Within a relatively short time after the system's origin, perhaps even before the major accretion phase of the terrestrial planets had been completed, life emerged on Earth.

The initially simple, microscopic organisms gradually became more complex and in the course of some 4 billion years evolved into an enormous variety of species, differing in size, habitats, and adaptive strategies. Most astonishingly, one of the species—Homo *sapiens* evolved language and culture, the ability to think and to reason, and the urge to embark on a systematic investigation of the Universe near and far.

The study of biological evolution on Earth will be the topic of the second part of this book. One generation passeth away, and another generation cometh; but the earth abideth forever. The sun also ariseth, and the sun goeth down, and hasteth to the place where he arose. The wind goeth toward the south, and turneth about unto the north; it whirleth about continually, and the wind returneth again according to his circuits. All the rivers run into the sea; yet the sea is not full; unto the place from whence the rivers come, thither they return again.

4

Origin of Life on Earth

Life began on Earth. There were no biological Big Bangs, nor extraterrestrial cultivators, just evolution from plausible nonliving beginnings." That is how John Scott (1981, 153), a biochemist at the Manchester Medical School in England, stated the basic assumptions scientists generally make about the origin of terrestrial life.

The question is, How did evolution from nonliving beginnings proceed? This chapter will attempt to answer this question, following theoretical developments that have gained widespread support from the scientific community since the late 1970s.

The story begins with the physical and chemical events that are believed to have taken place on the surface of the young Earth roughly 4 to 4½ billion years ago. At that time violent and nearly incessant volcanic eruptions occurred at many places, while at the same time the final accretion phase of the planet's formation drew to a close and its surface layers began to reach some semblance of equilibrium.

The eruptions spewed forth large quantities of water, carbon dioxide, molecular nitrogen, and many other molecules, from which the first atmosphere and the juvenile ocean formed. Driven by energy from sunlight, lightning, volcanic heat, and meteorite impacts, the inorganic molecules reacted chemically with each other and produced a great variety of organic molecules—amino acids, sugars, lipids, the bases of nucleic acids, and many more.

Gradually these molecules accumulated in the waters of the Earth until, in the words of John Haldane, "the primitive oceans reached the consistency of hot dilute soup". Today scientists believe that the temperature of the primitive ocean was probably close to the freezing point of water.

Furthermore, its content of organic molecules may not have been as concentrated as Haldane had envisioned it, at least not throughout most of its volume. Nevertheless, many people still refer to it as the "primordial soup." At present we are far from understanding all of the chemical reactions that took place on the young Earth.

It seems that simple organic molecules assembled into larger and more complex molecules similar to those found in today's organisms, as suggested by laboratory experiments. For instance, amino acids probably assembled into peptide chains, and nucleotides assembled into short strands of RNA and other kinds of nucleic acids. Surfaces of clays, lava, rocks, sand, and other readily available substances may have served as catalysts facilitating these assembly reactions.

According to the theory of the origin of life that I am presenting here, short strands of RNA were the first molecules in the primordial soup that carried information, albeit very little, and they are regarded as the starting point of the evolution toward cellular life. The short strands of RNA were capable of self-replication. In the process mistakes were made, so that the copied RNAs frequently differed from the original ones with regard to nucleotide sequence and length (recall that RNA nucleotides are of four types, with the bases A, G, C, and U).

Some of the RNA molecules were more successful than others in surviving and replicating themselves under the prevailing conditions. Thus, the two components that form the basis of the Darwinian theory of evolution - random creation of variation and natural selection—may have been introduced very early among the chemical reactions in the primordial soup.

As chemical evolution continued, different sets of RNA molecules coupled together into cooperative units. Some of the RNAs carried instructions for the assembly of primitive enzymes (peptide chains), while others acted as catalysts and contributed to the actual assembly of enzymes. Enzymes, in turn, helped in the replication of RNAs.

Eventually, some of the coupled units of RNAs and their enzymes became enclosed by membranes and the first primitive cells—the protocells—were born. Life had emerged from among the random and spontaneous chemical reactions in the primordial soup.

This brief summary outlines the key components of the events that many biologists and chemists believe may have been central to the origin of life on Earth, and the remainder of this chapter will fill in some of the details. At present this theory is based on many assumptions and contains many gaps.

It is not based on any direct evidence dating back to the primordial soup, for none has survived. Furthermore, laboratory experiments that attempt to simulate early Earth conditions and to reproduce the chemical reactions that took place then have been only partially successful in telling how the protocells arose.

Scientists think they understand in general terms how the Earth's atmosphere and ocean came into existence. They believe they know something about the formation of organic monomers from inorganic raw materials and the assembly of polymers from monomers. However, they still have only a very limited understanding of the emergence of order and information among the chemical reactions in the primordial soup.

And they know virtually nothing about the evolution from those initial chemical reactions to the formation of the first cells. The Nobel prize-winning German biochemist Manfred Eigen characterised very aptly our current state of knowledge of how life began: "Anyone attempting [to re-create life] would be seriously underestimating the complexity of prebiotic molecular evolution. Investigators know only how to play simple melodies on one or two instruments out of the huge orchestra that plays the symphony of evolution."

We should not be discouraged by this lack of knowledge. Let us accept the problem of the origin of life as one of the great challenges facing science today. Let us also accept the fact that time inevitably diminishes and sometimes erases evidence of long-ago events. Hence, our first task is to discover the fragments of evidence that have survived. Our second task is to make good use of them.

That is how Darwin and Hubble confronted their scientific challenges, which also dealt with events of long ago and for which much of the original evidence had been erased by time. Darwin deduced his theory of the origin and evolution of the species mainly from data gathered on a single trip around the globe, and Hubble based his proposal about the expansion of the Universe on measurements of recession velocities of about two dozen distant galaxies.

Thus, there are precedents in the history of science that fragmentary information is no barrier to the development of feasible theories. Let us be optimistic that this will also be true of the current scientific attempts to reconstruct the events that led to life on Earth.

GEOLOGIC ACTIVITY ON THE YOUNG EARTH

When the Earth was formed roughly 4.5 billion years ago, it was a hot, partially molten mass without an ocean or much of an atmosphere. Most of the heat came from gravitational energy that was released when planetesimals collided and fell together to form the Earth, as discussed in chapter elsewhere in this chapter.

Additional gravitational energy was released when the high-density iron and nickel of the proto-Earth sank toward the center to become the core of our planet, and lighter rocky material rose toward the surface to form the mantle and crust. Some of the energy also came from the decay of radioactive isotopes.

These energies were liberated much more rapidly than they could be radiated away, and consequently they accumulated as heat. Only a fraction of this heat has been lost during the intervening eons. Even today, our planet's central temperature is still approximately 4300 K.

During the final accretion phase, Earth acquired a surface layer of low-density rocks rich in many kinds of volatiles including water, carbon dioxide, molecular nitrogen, and organic compounds. This rocky material was probably derived from carbonaceous chondrites, which bombard the Earth to this day, along with other types of meteorites, though at a much reduced rate.

The outermost layers of the Earth radiated their heat into space and cooled to the point where they crystallised and hardened. They became the basaltic and granitic rock layers that form the crust of the Earth and float like rafts on the denser underlying mantle. The crust is primarily responsible for maintaining the Earth's high internal temperature.

It has a very low thermal conductivity, which slows the rate of heat flow from the Earth's interior to the surface. This can be seen in some of the desert caves in the western United States where the snow and ice that drift in during the winter stay throughout the hot summer months, even though they are separated from the surface by only a few meters of rock.

Another feature that contributes to the maintenance of the Earth's high internal temperature is the presence of long-lived radioactive elements - uranium-235 and -238, thorium-232, and potassium-40-in the crust. The decay of these elements steadily releases heat and is the source of much of the geothermal energy that flows to the Earth's surface. Thus, the crust, enveloping the Earth, acts like an electric blanket: Its low thermal conductivity corresponds to the insulating qualities of the wool or polyester, and its radioactivity corresponds to the electric heat output of the blanket.

One consequence of the Earth's high interior temperature is that the outer part of the core and the mantle have never hardened into rigid structures, but have remained in molten or "pasty" states, resembling fluids of high viscosity.

Another consequence is that powerful convective currents are generated in the mantle, which relentlessly push and pull on the overlying layers (that is, on the *lithosphere*) and prevent them from settling into a permanent configuration. That is why our planet's surface features continuously change over geologic time.

Continents converge and break up, ocean basins come and go, and mountain ranges are lifted up and weathered down. The pushing and pulling on the Earth's lithosphere by currents in the underlying mantle involve enormous forces and energies.

Usually we are not aware of those geologic activities because they happen so slowly; but earthquakes, volcanic eruptions, geysers, and hot springs are reminders that we live on a restless and dynamic planet. This restlessness must have been much more severe when the Earth was first formed than it is today.

The Earth's interior was hotter then and its temperature had not yet had time to adjust to a smooth gradient from the center to the surface. Consequently, the currents in the mantle must have been stronger. The lithosphere was still crystallizing and had not yet achieved its present thickness and rigidity. This crystallizing was slowed by the steady release of heat from the decay of the radioactive elements, which were initially much more abundant than they are today.

Because it was thinner and less rigid than it is now, the lithosphere of the young Earth was more easily deformed by the mantle currents than it is today. As a result, earthquakes and volcanic eruptions, accompanied by huge lava flows, must have occurred almost incessantly and with great intensity over large areas of the young planet's surface, much as they occurred on the Moon, Mercury, Mars, and, perhaps, on all planetary bodies of intermediate size.

In addition to earthquakes and volcanism, which are processes created by conditions within the Earth, there also was violence from outside. When our planet was formed and had reached approximately its final mass and size, there was still plenty of interplanetary debris—planetesimals, comets, rocks, and dust—left from the original protoplanetary disk, as discussed at some length in chapter elsewhere in this chpater. For hundreds of millions of years this debris kept falling onto the Earth at a high rate until most of it had been swept up.

Even today some traces remain, as indicated by the roughly 30 tons of matter that fall onto Earth every day in the form of "shooting stars" and meteorites. The largest of the planetesimals that bombarded the young Earth probably weighed many billions of tons and were comparable in size to the asteroids that still orbit the Sun today.

On impact, they shattered the crust, carved out huge impact craters, and threw molten and pulverised crustal material across the planet's surface.

Table 4.1 : Partial listing of Molecules.

CH_4	methane	H_2O	water
CO	carbon monoxide	H_2S	hydrogen sulfide
CO_2	carbon dioxide	NH_3	ammonia
CO_3	carbonate	NO_3	nitrate
C_2H_4	ethylene	N_2	molecular nitrogen
C_2H_6	ethane	O_2	molecular oxygen
HCN	hydrogen cyanide	PO_4	phosphate
H_2	molecular hydrogen	SO_4	sulfate
H_2CO	formaldehyde		

ORIGIN OF THE EARTH'S ATMOSPHERE AND OCEAN

The earthquakes, volcanic outbursts, and bombardments by meteorites, which ravaged the young Earth so regularly, did not just churn the crust and produce large lava flows. As hot lava reached the surface and meteorites heated their impact areas to incandescence, gases that had been trapped in the rocks burst into the open in huge amounts—gases of water (H_2O), carbon dioxide (CO_2), molecular nitrogen (N_2) and, in lesser amounts, of molecular hydrogen (H_2), carbon monoxide (CO), methane (CH_4), ammonia (NH_3), hydrogen sulfide (H_2S), and many others. Our planet was acquiring its first atmosphere.

This kind of outgassing (releasing of gases) can still be observed today, although at a considerably diminished rate, in the hot springs and geysers of Yellowstone National Park, active volcanoes such as Mount St. Helens, and many other places of geothermal activity. For example, volcanic eruptions are usually accompanied by the emission of thick and often foul—smelling clouds of gases that billow for miles into the atmosphere. The gases originate from within the lava. They escape into

Table 4.2 : Relative Abundances (in Percent) by Mass of the Most Common Elements in the Sun, Entire Earth, Earth's Crust, and Human Body.

Element	*Suna Earthb*	*Entire Crustb*	*Earth's*	*Human body*	
Hydrogen	73.6	—	—	10.	(63.)
Helium	24.8	—	—	—	—
Carbon	0.29	—	—	18.	(9.5)
Nitrogen	0.093	—	—	3.1	(1.4)
Oxygen	0.77	30.	46.	65.	(25.)
Neon	0.12	—	—	—	—
Sodium	0.0029	—	2.1	0.1	(0.03)
Magnesium	0.046	13.	4.	0.04	(0.01)
Aluminum	0.0049	1.1	8.	—	—
Silicon	0.069	15.	28.	—	—
Phosphorus	0.0007	—	—	1.	(0.2)
Sulfur	0.038	1.9	—	0.3	(0.05)
Chlorine	0.0011	—	—	0.2	(0.03)
Argon	0.018	—	—	—	—
Potassium	0.0003	—	2.3	0.4	(0.06)
Calcium	0.0057	1.1	2.4	2.	(0.3)
Iron	0.16	35.	6.	—	—
Nickel	0.0084	2.4	—	—	—

Note: The last column, given in parentheses, refers to abundance by number relative to a total of 100.

the open when the hot lava reaches the surface of the Earth and is no longer subjected to the high pressures deep below the ground. Quite often the gases bubble forth in ways that give the resulting rocks a frothy appearance and make them lighter than water. During some eruptions the gases burst forth so violently that they shatter the lava into fine-grained dust known as ash, which may be thrown hundreds or even thousands of kilometers into the surrounding areas.

Water

Despite the heat that accompanied outgassing on the early Earth, the average temperature of the atmosphere was probably not far above the freezing point of water, for the energy output of the Sun—our planet's major source of heat—was then only about two-thirds of what it is today.

Hence, the water, which was the most abundant molecular species emitted by volcanism and meteorite impacts, did not remain in gaseous form for long. It condensed into rain or snow and fell to Earth. Very likely the early rains came down in torrents, accompanied by storms, lightning, and thunder.

The first rivers began to flow, glaciers and ice caps accumulated in the polar regions, and low-lying basins filled with liquid water to become our planet's juvenile ocean.

Carbon Dioxide

Volcanic outgassings on the early Earth poured forth not only water but also many other gases. The most abundant gas after water was carbon dioxide. This presented a potential danger for the eventual origin and evolution of life on Earth because a planetary atmosphere that contains substantial amounts of CO_2 heats up to high temperatures by the *greenhouse effect*.

This effect comes about as follows. Solar radiation, whose wavelengths correspond mainly to the colours blue through red, penetrates relatively easily through the atmosphere and is absorbed by the ground. The absorbed energy heats the ground, which reradiates it in the infrared part of the spectrum. Infrared (IR) radiation has long wavelengths and, unlike the original solar radiation, much of it becomes absorbed by the CO_2 in the atmosphere.

The CO_2 molecules reradiate this absorbed radiation, with some of it being sent back toward the ground. The net result is that the IR radiation does not readily escape into space but becomes trapped in the atmosphere, heating it as well as the ground.

Fortunately, the greenhouse effect had only a small effect on the young Earth's surface temperature because the rains washed the CO_2 out of the atmosphere before it could accumulate to dangerous levels. This happened because CO_2 dissolves in liquid water (but not in gaseous water) and forms carbonic acid (H_2CO_3):

$$CO_2 + H_2O \text{ (liquid)} \rightarrow H_2CO_3.$$

The rains and rivers brought the carbonic acid in contact with the ground, where it leached positive ions of calcium (Ca^{++}) and magnesium (Mg^{++}) out of rocks and combined with them to form limestone ($CaCO_3$)

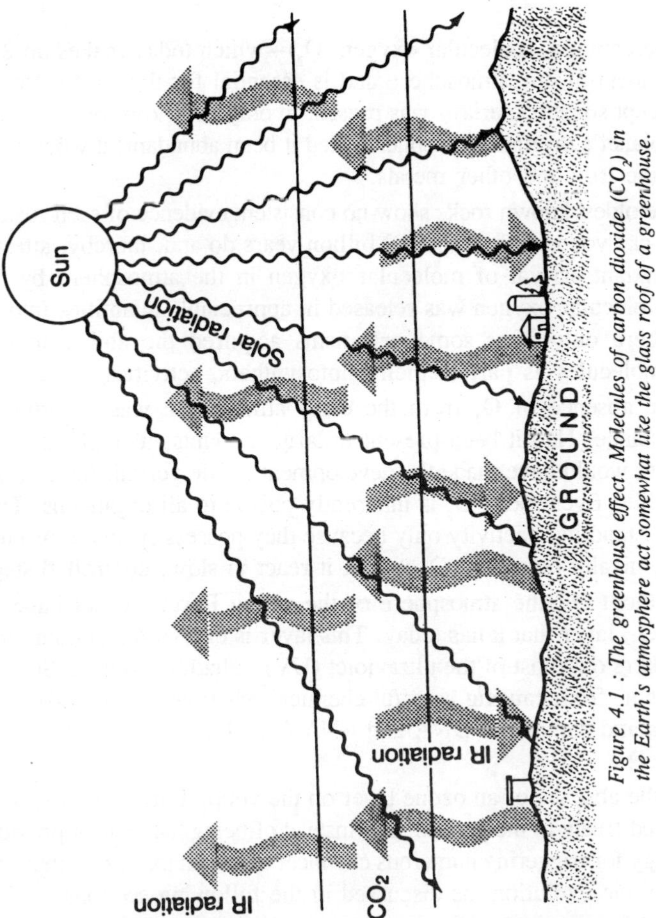

Figure 4.1 : The greenhouse effect. Molecules of carbon dioxide (CO_2) in the Earth's atmosphere act somewhat like the glass roof of a greenhouse.

and dolomite [$CaMg(CO_3)_2$]. The limestone and dolomite then became deposited as sediments on the ocean floor. Thus, most of the original CO_2 that had been released into the atmosphere became locked up in sedimentary rocks. (Note that after shale and sandstone, limestone constitutes the most abundant sedimentary rock type of the present-day terrestrial crust.)

The third most abundant gas released by volcanism and meteorite impacts on the early Earth was molecular nitrogen, N_2. It is chemically rather inert and did not condense into liquid form at the temperatures that prevailed on Earth, nor did it combine with water. It stayed in the atmosphere and, in the course of time, accumulated to become its dominant molecular constituent.

Oxygen

Interestingly, molecular oxygen (O_2)—which today makes up 21% (by number) of our atmosphere and is essential for the survival of all life (except some bacteria)—was present in only very low concentrations in our planet's original atmosphere. Had it been abundant, it would have rusted the iron and other metals.

The oldest known rocks show no consistent evidence of such rusting. Only rocks younger than about 2 billion years do and, thereby, attest to the abundant arrival of molecular oxygen in the atmosphere by that time. Molecular oxygen was released in appreciable quantities into the atmosphere only after some organisms acquired the ability to split water molecules as part of their photosynthetic activities.

The absence of O_2 from the early atmosphere was a fortuitous circumstance. Had it been present in large amounts, its high chemical reactivity would have made the development of life very difficult, if not impossible. Even today O_2 is inherently poison to all organisms. They can withstand its reactivity only because they possess special molecules that chemically bind with O_2 and let it react in slow, controlled steps.

Without O_2, the atmosphere of the young Earth did not have the ozone (O_3) layer that it has today. This layer is crucial for our survival, for it filters out most of the ultraviolet (UV) radiation from the Sun and prevents it' from causing harmful chemical reactions in our cells, such as burns and damage to DNA and RNA (which may lead to cell death or cancer).

In the absence of an ozone layer on the young Earth, UV radiation penetrated freely to the ground. But instead of being harmful, it provided the energy for triggering numerous chemical reactions that were important for prebiotic evolution, as discussed in the following sections.

The Next Step

Very likely, the first outgassings on the primitive Earth began long before the planet was fully formed and they continued with great intensity for hundreds of millions of years thereafter. Only when the initial burst of earthquake activity, volcanism, and meteorite impacts had abated, approximately 4.0 to 3.5 billion years ago, did the outgassing slow down.

Geologic evidence indicates that by then the oceans covered large areas of our planet, and we may assume that the atmosphere had acquired a substantial fraction of its present content of N_2. Most of the outgassed CO_2 had been washed out of the atmosphere and locked up in sedimentary rocks, but not all of it. Some of the CO_2 together with other gases that had *been* discharged in small amounts, such as H_2, CH_4, NH_3, and H_2S, embarked upon a rather different and much more interesting evolutionary course. They reacted chemically with each other to form molecular structures of ever-increasing complexity.

ORGANIC CHEMISTRY ON THE EARLY EARTH, PHASE ONE: SYNTHESIS OF MONOMERS

There are good reasons to believe that conditions on the surface of the young Earth were quite suitable for the occurrence of a great variety of chemical reactions and the production of many kinds of molecules. Volcanic outgassings supplied ample amounts of atomic and molecular raw materials.

The temperature was low, but not so low as to freeze water everywhere. Lightning, UV radiation from the Sun, geothermal heat from volcanoes and hot springs, and meteorite impacts supplied plenty of energy for driving the reactions. At the same time, the ocean and ground offered protection from too much UV radiation and heat that might have destroyed the molecular products.

Among the molecular raw materials on the surface of the young Earth, H_2O, N_2, and CO_2 were the most abundant. They offered a nearly inexhaustible supply of hydrogen, carbon, nitrogen, and oxygen. Hence, organic molecules must have been among the dominant products of the chemical reactions.

This is the case in interstellar clouds, in which these elements are also abundant and in which organic molecules are manufactured copiously. The same was true of the Sun's protoplanetary disk, as manifested by the presence of organic molecules in meteorites.

Just how readily organic molecules form under conditions similar to those on the early Earth was first demonstrated experimentally by two American chemists, Stanley L. Miller and Harold C. Urey, in the early 1950s, using a glass apparatus as shown in figure elsewhere in this chapter. They filled the lower part of the apparatus, which included a small flask, with water to simulate the juvenile ocean.

They pumped a gaseous mixture of CH_4, NH_3, and H_2 into the upper part to represent the primitive atmosphere. Then they boiled the water in the small flask to produce steam and to drive the gases in a closed circuit through the apparatus.

At the same time, they generated electric sparks in the larger upper flask to simulate lightning in the primitive atmosphere and to provide a source of energy for chemical reactions. Below the large

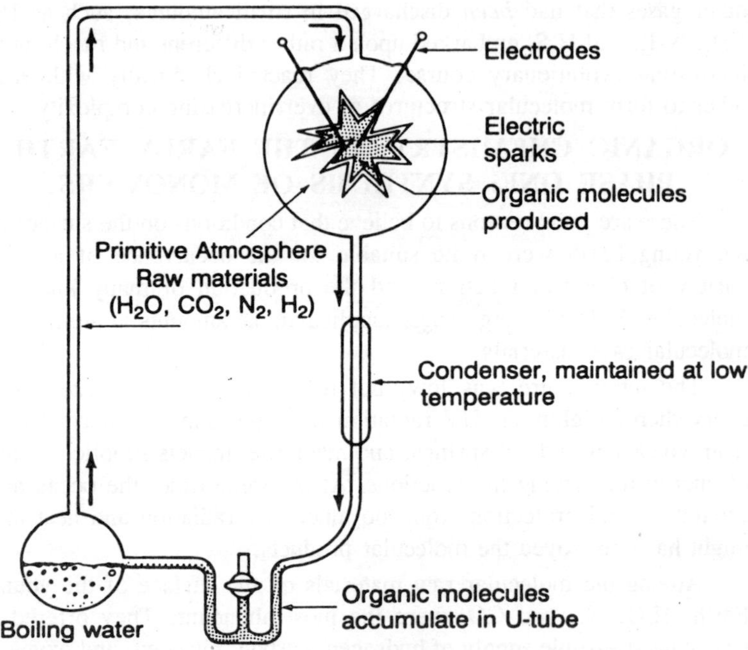

Figure 4.2 : Apparatus used in Miller-Urey experiments.

flask, they cooled the circulating gases so that the water condensed and returned as droplets to the lower part of the apparatus, thus completing the circuit. After they had run the experiment for a day or two, the water in the lower part of the apparatus turned pink and later it became red.

This indicated, and careful analysis confirmed, that organic molecules (such as hydrogen cyanide, formaldehyde, sugars, amino acids, bases) had been formed by the electric sparks and were trapped in the liquid water. When Miller and Urey first carried out their experiment, it was commonly thought that hydrogen-rich molecules had been the major constituents of the Earth's primitive atmosphere.

Hence, they used the gases CH_4 and NH_3 instead of CO_2 and N_2. Since then, the experiment has been repeated many times with various combinations of CH_4, C_2H_4 (ethylene), C_2H_6 (ethane), CO_2, CO, NH_3, N_2, H_2S, H_2, and H_2O and with different sources of energy, from intense heat (from about 1100 to 1600 K) to UV radiation and electric sparks.

Some experimenters added sand, clay, or lava rocks to simulate the catalytic effects the ground of the young Earth might have had on the chemical reactions. All of the experiments yielded rich mixtures of organic molecules.

Apparently, the main requirement for synthesizing organic molecules is the presence of hydrogen, carbon, nitrogen, and oxygen in some molecular form or other. The details -whether the atomic raw materials are supplied as H_2O, CO_2, N_2, and H_2, as H_2O, CH_4, NH_3, and H_2, or in some other form—matter little. Nor does it matter much what kind of energy source is used, as long as it is sufficiently intense to drive the chemical reactions.

The chief limitations are that oxygen not be present as $_{02}$ and that the newly formed molecules do not remain exposed to the energy sources for too long. If they are, the same energy source that creates them also destroys them. Protection from too much energy is achieved by dissolving the molecular products in liquid water (that is, in the juvenile ocean) or by allowing them to attach themselves to the surfaces of grains of sand, clay, mud, or lava rocks.

Note, however, that the Miller-Urey experiments do not yield many organic molecules of prebiotic importance if the only raw materials are H_2O, CO_2, and N_2. To enhance the production of organic molecules, H_2, CH_4, NH_3, or other hydrogen-rich molecules need to be present among the starting materials, at least in small amounts (without them, the chemical reactions are dominated by the reactivity of oxygen).

This is particularly true of the production of hydrogen cyanide (HCN), which is a precursor molecule of many of the more complex organic molecules that are believed to have played significant roles in prebiotic chemistry.

Synthesis of Amino Acids, Sugars, and Bases

How are the simple starting materials converted into organic molecules in experiments simulating primitive Earth environments? Much attention has been paid recently to answering this question. Here are a few examples of these chemical reactions.

When molecules such as H_2O, CO_2, N_2, and H_2 are exposed to electric sparks, intense heat, or UV radiation, their bonds are broken and, temporarily, free atoms (H, C, N, O) and molecular fragments (for example, OH and CO) result.

Almost instantaneously, new bonds reform between the atoms and the molecular fragments, but often in combinations that are different from the original ones. This breaking and reforming of chemical bonds constitutes the mechanism of chemical reactions by which new kinds of molecules are created. Such chemical reactions take place in statistically predictable ways and according to well-known physical laws. Given identical starting conditions, the molecules produced *will,* on the average, always turn out the same.

Two of the most common molecules that result during the initial stages of a Miller-Urey experiment are hydrogen cyanide (HCN) and formaldehyde (H_2CO). Both of them are important intermediates in the formation of still more complex organic molecules, in particular of amino acids.

For instance, reactions among HCN, H_2CO, and water yield the amino acid glycine. The reactions involve several steps, but they may be summarised by this chemical equation:

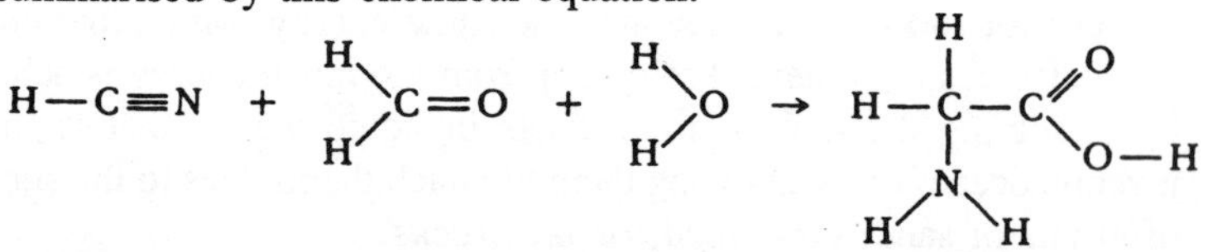

$$H{-}C{\equiv}N \;+\; (H)_2C{=}O \;+\; (H)_2O \;\rightarrow\; H{-}C(H)(NH_2){-}C({=}O){-}O{-}H$$

Hydrogen cyanide Formaldehyde Water Glycine

Reactions involving more complex aldehydes than formaldehyde yield more complex amino acids. The nature of the end product depends somewhat on the sources of energy used and on the presence of certain atoms among the starting materials.

If electric sparks are used, *glycine*, *alanine*, *leucine*, *serine*, *threonine*, *asparagine*, and other relatively simple amino acids result. If the energy source is heat of about 1600 K, some of the products are amino acids with ring structures, such as phenylalanine and tyrosine. If UV radiation is used and H2S is present in the gas mixture, small amounts of sulfur-containing amino acids are formed.

In addition to being precursor molecules for the making of amino acids, formaldehyde and hydrogen cyanide may also be the starting materials for the synthesis of other organic molecules. For instance, reactions among five molecules of formaldehyde, in the presence of calcium carbonate, produce a complex mixture of end products, including the five-carbon sugar *ribose:*

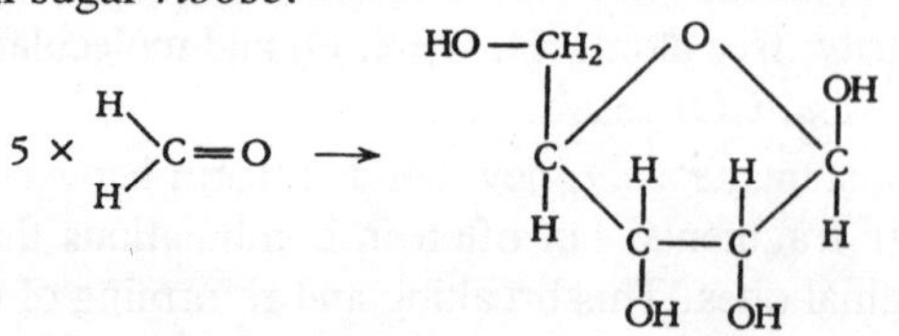

$$5 \times (H)_2C{=}O \;\rightarrow\; \text{Ribose}$$

Formaldehyde (×5) Ribose

As before, this formula merely summarizes the input material and the final product of the reaction. The actual reaction sequence is much more complex, involving several steps. In the first step, two formaldehydes react to form glycolaldehyde:

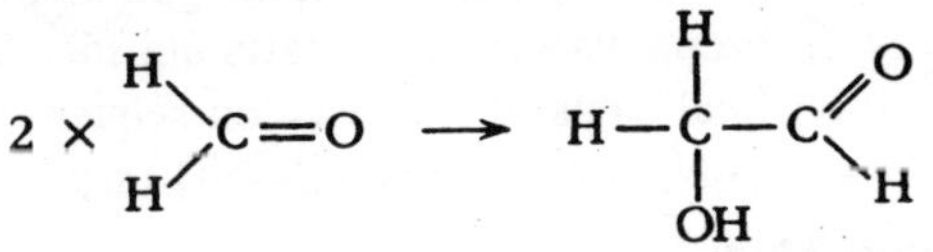

$$2 \times \mathrm{H_2C{=}O} \longrightarrow \mathrm{H{-}CH(OH){-}CHO}$$

Formaldehyde (×2) Glycolaldehyde

Glycolaldehyde may be either the precursor of ribose, if three more formaldehydes are added (as in the reaction shown above), or the precursor of the amino acid serine, if HCN and H2O are added. In either case, the reactions are *autocatalytic*. At first very little happens.

Then suddenly, after several hours, ribose and serine are produced. Apparently, molecules of glycolaldehyde slowly form during the induction period. Once they exist, they act as catalysts for the production of more glycolaldehyde and then the reactions to ribose and serine run their course rather rapidly. Six-carbon sugars, such as glucose and fructose, are formed by similar reaction pathways.

Not all organic molecules of biological interest are as readily produced as the amino acids and sugars. For example, the syntheses of the bases of DNA and RNA call for rather high concentrations of starting molecules, which on the early Earth probably occurred only under unusual circumstances.

Adenine, which is the easiest base to make (perhaps that is why it is the most common base found in organisms, being present in RNA, DNA, and ATP), requires the reaction of five molecules of hydrogen cyanide:

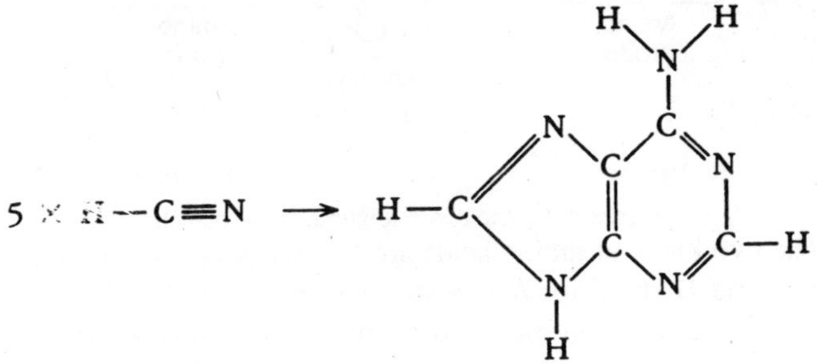

$$5 \times \mathrm{H{-}C{\equiv}N} \longrightarrow \text{Adenine}$$

Hydrogen cyanide (×5) Adenine

Again, this formula is merely a summary. In reality, several sequential reactions must occur, and the presence of UV radiation and ammonia are helpful. *Guanine,* the other double-ringed base, is very similar to adenine except that it possesses an oxygen atom and its side groups are slightly different. Its synthesis starts out identically to that of adenine, but in the final steps of the reaction sequence, water and cyanogen (N=—C—C=—N) or urea (H_2N—C—NH_2) need to be present.

L- and D -Amino Acids

These have been just a few examples of the numerous chemical reactions that occur in Miller-Urey type and other kinds of experiments simulating prebiotic terrestrial conditions. Many of the molecules produced are identical to those found commonly in present-day life. Others are rather novel and have unusual structures.

For instance, many of the amino acids have two configurations, one being the mirror image of the other. In one configuration, the side chain points to the *left* as one looks from the carboxyl group to the amino group (for details, see the Proteins section in the Introduction to Part Two). In the other, the side chain points to the *right*. The two configurations are illustrated in Figure elsewhere in this chapter for alanine.

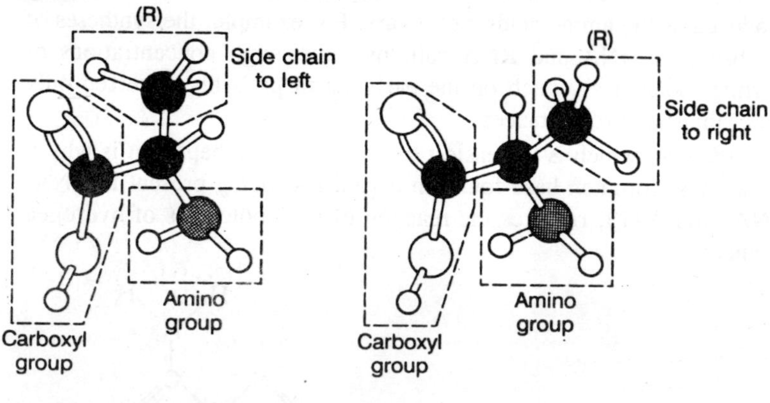

Figure 4.3 : The two configurations of alanine.

The two kinds of amino acids are known as L- and D-amino acids, where L and D stand for *levo* and *dextro* meaning "left" and "right." Both kinds are found in meteorites, and it is reasonable to assume that they were manufactured by prebiotic terrestrial chemistry as well. In contrast, with few exceptions today's organisms produce only L-amino acids. Somehow, during the origin of life, L-amino acids were selected

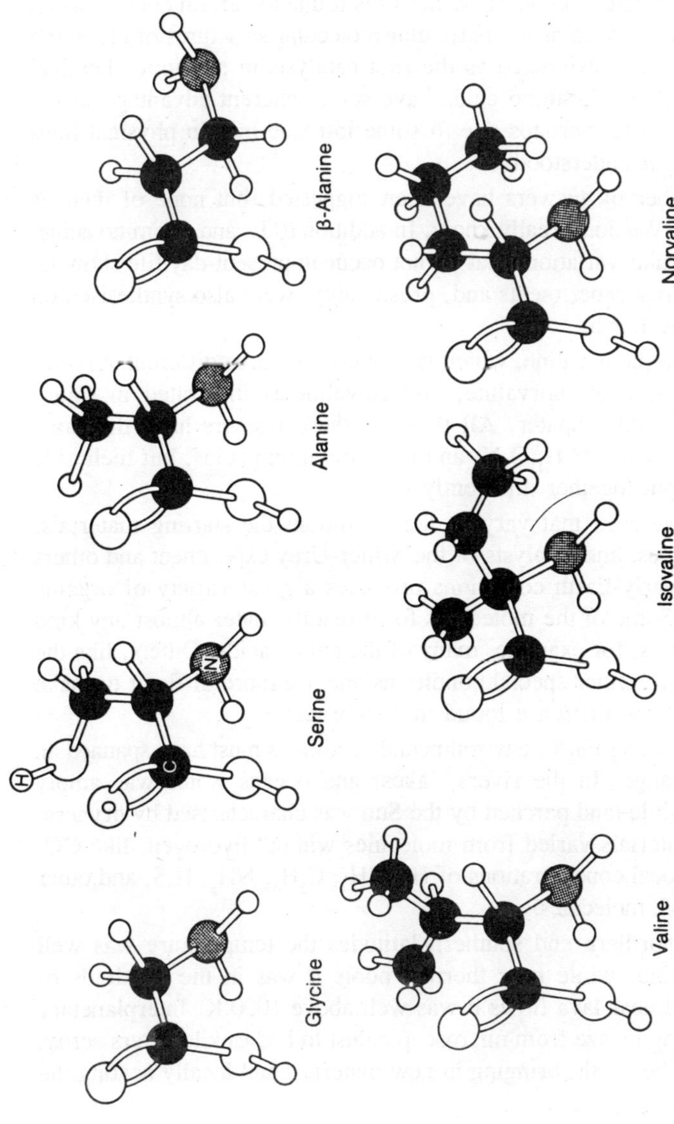

Figure 4.4 : Examples of amino acids produced on the early Earth. Many of the amino acids existed in several isomeric forms and, with the exception of glycine, they occurred as both L- and D-types.

over D-amino acids. Once that had happened, enzyme-aided biochemical reactions kept manufacturing chiefly the L-types.

Was this selection an accident? Was it due to certain conditions on the early Earth, such as the particular molecular structures of clays and rocks, that may have acted as the first catalysts in prebiotic chemical reactions? Or do L-amino acids have some inherent advantage in the chemistry of life, perhaps due to some intrinsic bias in physical laws that is not yet understood?

A number of answers have been suggested, but none of them is convincing. We don't really know. In addition to L- and D-amino acids, other molecular variations that do not occur in present-day life show up in Miller-Urey experiments and, presumably, were also synthesised on the prebiotic Earth.

An example is valine, which is produced in three different versions or *isomers*—valine, norvaline, and isovaline-as illustrated in figure elsewhere in this chpater. All three of these isomers have the same chemical formula, $C_5H_{11}O_2N$, and they are amino acids, but their side chains are put together differently.

We have seen that varying the conditions-the starting materials, energy sources, and catalysts-of the Miller-Urey experiment and others simulating early-Earth conditions produces a great variety of organic molecules. Some of the molecules form readily under almost any kind of condition as, for example, many of the amino acids. Others, like the bases, require rather special conditions and are more difficult to make (though some of them are found in meteorites).

On the young Earth, environmental conditions must have spanned an enormous range. In the rivers, lakes, and oceans water was amply available, while land parched by the Sun was characterised by dryness. The raw materials varied from molecules without hydrogen, like CO_2 and N_2, to local concentrations of H_2, CH_4, C_2H_6, NH_3, H_2S, and other hydrogen-rich molecules.

In the northern and southern latitudes the temperature was well below freezing, while near thermal pools it was in the hundreds of degrees, and near lava flows it was well above 1000 K. Interplanetary debris ranging in size from microscopic dust to bodies kilometers across bombarded the Earth, bringing in new materials and locally heating the atmosphere and ground.

The Sun's UV radiation penetrated freely to the ground. Primordial storms produced lightning and released concentrated forms of heat. Rocks and mineral grains of various sizes, textures, and compositions offered additional raw materials, besides those discharged by outgassing, and were available to serve as catalysts for many of the reactions.

From day to night, season to season, and year to year, the early terrestrial conditions fluctuated. We may expect that in the course of thousands to many millions of years a broad range of chemical reactions occurred and that organic molecules were produced in great abundance and in many different forms.

Even molecules that are very difficult to make were produced occasionally. The American biochemist, George Wald, put it as follows: "Given so much time, the 'impossible' becomes possible, the possible probable, and the probable virtually certain. . . . Time is in fact the hero of the plot."

Once the molecules were formed, they were not easily destroyed as they would be today by bacteria and molecular oxygen. The oceans, rivers, and lakes offered protection from UV radiation and excessive heat as did sand, clay, and mud. Thus, organic molecules accumulated until, as Haldane suggested, the primordial ocean reached the consistency of dilute soup.

We do not know exactly how concentrated the soup became. That depended on the rate of synthesis of organic molecules as well as on the rate of their destruction. It also depended on the volume of the ancient ocean. However, we may be certain that at least locally, such as in shallow marine basins, tide pools, and lakes, its concentration was sufficiently great for the next phase in chemical evolution to get started—the assembly of simple organic molecules (the monomers) into polymers.

ORGANIC CHEMISTRY ON THE EARLY EARTH, PHASE TWO: SYNTHESIS OF POLYMERS

The prebiotic synthesis of simple organic molecules was only the beginning of a complex and multifaceted evolution of chemical reactions. As amino acids, sugars, bases, and other molecules accumulated and their concentrations in the primordial soup and in the clay, sand, and mud of the young Earth increased, further chemical reactions linked them together into larger molecular structures, namely polymers.

Amino acids were joined into peptide chains; ribose, bases, and phosphates were combined into nucleotides; and nucleotides were connected into strands of RNA and, possibly, other kinds of nucleic acids. These are just a few examples of polymers that probably were constructed from monomers by prebiotic chemistry.

Many early polymers were identical or similar to those found in

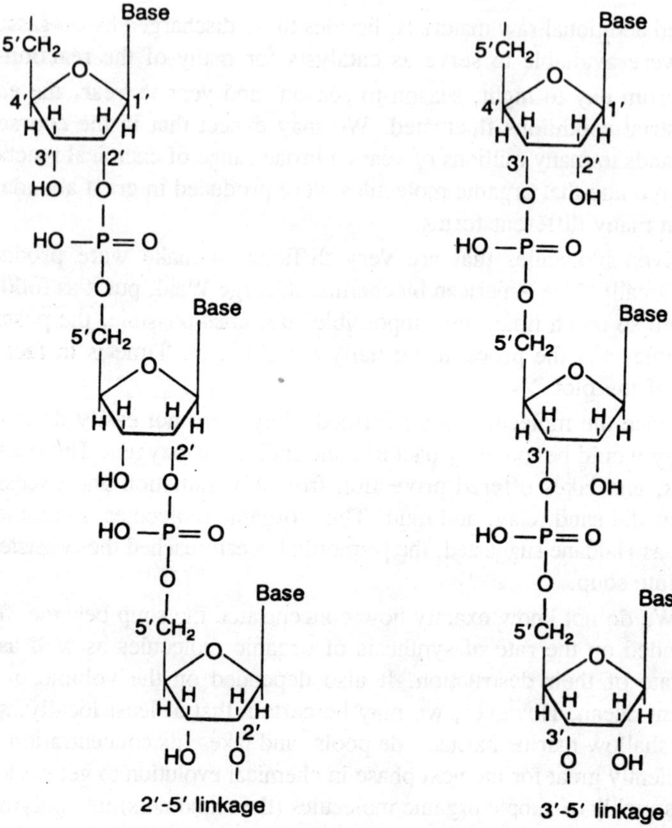

Figure 4.5 : Chemically possible RNA "backbone" structures. The backbone structure of RNA, consisting of ribose sugars linked by phosphates, can exist in two forms: 2'-5' linkages and 3'-5' linkages.

nature today, but others were quite different, just as some of the monomers were different. For example, it is very likely that nucleotides were bonded together into short RNA and DNA polymers by both the so-called 3'-5' linkage, which is found in contemporary cells, and by the 2'-5' linkage.

The chemical reactions that produced polymers from monomers on the early Earth were probably the same dehydration reactions that take place in today's organisms (though initially they were not catalyzed by enzymes): A hydrogen atom was removed from one monomer and an OH fragment from another, and then a chemical bond was formed between the two monomers.

The H and OH were combined into a molecule of H_2O, which was released into the environment (see Introduction to Part Two). In today's organisms the assembly of monomers into polymers takes place in the interior of cells. There the monomers are well concentrated, so that collisions and, hence, reactions among them are frequent; the energy required for tearing the Hs and OHs from the monomers is amply supplied by ATPs; and enzymes are present to help remove the water molecules and speed up the formation of the bonds.

On the prebiotic Earth conditions for making polymers were considerably less favourable. Because cells did not yet exist, reactions took place in the environment at large. There the concentrations of monomers were generally low compared to the interiors of present-day cells; energy was available, but most of it not in forms best suited for driving the dehydration reactions; and enzymes were lacking.

In the oceans and lakes the absence of enzymes presented a particularly serious problem. Without enzymes it was difficult to form bonds by removing water molecules from monomers and expelling them into surroundings that already consisted largely of water. The reactions were much more likely to run in the opposite direction, namely in the direction of taking water molecules from the surroundings, adding them to polymers, and splitting their bonds (a process called *hydrolysis,* see Lipids section, Introduction to Part Two).

Concentration Mechanisms

With all of these obstacles, how were the first polymers assembled? We may assume that in local areas conditions were occasionally suitable for polymerisation. For instance, evaporation of water in shallow lakes and ponds during dry spells might have satisfied the requirement of concentrating the monomers by leaving most of the organic molecules behind on the muddy bottoms.

The heat near hot springs or from freshly expelled lava flowing into bodies of water would have had similar effects. The periodic rising and falling of the water level in tide pools would have regularly concentrated the molecules, especially in hot, dry climates. Still another concentration mechanism would have been the freezing of lakes and ponds during winter.

As water froze to ice, organic molecules would have become concentrated in the remaining liquid water. (The same method was used by American pioneers in the making of applejack. A barrel of cider was put outside during the freezing weather and left standing until most of the water in the cider was solid ice, leaving the liquid alcohol concentrated as applejack in a small volume near the center of the barrel. The colder the temperature, the higher was the proof and, hence, the alcohol concentration of the applejack.)

The concentration mechanisms just discussed were all a result of changes in the environment. Concentrations may also have been accomplished by the organic molecules themselves. For instance, in the laboratory when amino acids are put in water that is then heated to above the boiling point, peptides form and cluster spontaneously into spherical, membrane-enclosed structures called *proteinoid microspheres*.

These structures are about 1 μm across and, during their self-assembly, concentrate in their interiors many of the organic molecules present in their vicinity. Another concentration mechanism occurs when certain combinations of organic polymers, such as peptides, carbohydrates, and nucleic acids, are mixed together in water.

They spontaneously organize themselves into clusters that are marked off from their environment by structured layers of water molecules. These clusters are called *coacervates*. They have sizes of up to 500 pm and, like the proteinoid microspheres, concentrate organic molecules in their interiors. Both proteinoid microspheres and coacervates may well have played significant roles in concentrating organic molecules in the primordial soup of our planet.

Energy Sources

The second requirement for assembling polymers - the availability of energywas probably at first satisfied by lightning, heat, and UV radiation, even though these sources of energy are in general not very efficient in linking monomers together. As the number and diversity of monomers increased, the chemical energy stored in the bonds of some of them became available for dehydration reactions as well.

Hydrogen cyanide, cyanogen, and other nitrogen-containing molecules similar to cyanogen were probably most effective in driving the reactions. Very likely, chains of phosphate molecules, the *polyphosphates,* were also important energy sources. In fact, polyphosphates may have been the precursors of ATP, which consists of a triphosphate chain attached to an adenine-ribose trunk.

Polyphosphates must have been quite abundant on the surface of the early Earth. They would have formed readily by mild heating of minerals containing phosphate, particularly during dry periods when the minerals became concentrated (along with other materials, such as organic molecules) on the bottoms of lakes and ponds.

Catalysts

The third requirement for making polymers is the presence of catalysts for removing water molecules from the monomers and speeding up the reactions. Until enzymes evolved, the most obvious catalysts were the surfaces of mineral grains. Many clays consist of very thin sheets of silicates, that are separated from each other by molecules of water.

The water layers would have given the organic monomers easy access to the silicate surfaces, where they could have become attached and been ready to undergo dehydration reactions with newly arriving monomers. Polyphosphates would have been amply available in the minerals as sources of energy.

Experiments indicate that peptide chains as long as one hundred amino acids form in the presence of certain clays. The assumption that the surfaces of mineral grains acted as catalysts on the prebiotic Earth is supported by evidence from astrophysics. It appears that many of the molecules (including organic molecules) that are observed in the dark gas and dust clouds of spiral and irregular galaxies are also assembled on the surfaces of dust grains.

Experiments carried out with proteinoid microspheres suggest that they and other accidentally produced protein-like structures might have acted as catalysts as well. However, until the development of genetic information, the sequences in which amino acids were assembled into peptides and protein-like structures remained largely a matter of chance. Therefore, whatever catalytic properties such protein-like structures may have possessed lacked the specificity and efficiency that characterize enzymes in contemporary cells.

This section described some of the processes by which organic monomers are thought to have been assembled into polymers on the early Earth. However, at present scientists are far from understanding the full range of chemical reactions that took place then. No doubt, processes other than those described played important roles in the assembly of polymers as well.

Our knowledge is limited because it is very difficult to duplicate in the laboratory the early terrestrial environment. As noted in the previous section, there are a great many variables to consider: an enormous variety of chemically possible precursor molecules, a broad range of likely concentration mechanisms, fluctuations in temperature of the environment from below freezing to well above 1000 K, numerous sources of energy, and the effects of many different kinds of inorganic materials from clays to rocks, sand, mud, lava, and sediments. Furthermore, experiments of prebiotic chemical evolution must be conducted in closed apparatuses in order to avoid contamination by microorganisms. In contrast, chemical evolution on the early Earth occurred in an open and ever-changing environment consisting of the atmosphere, water, and land.

Above all, experiments are limited by the factor of time. Prebiotic chemical evolution took place over thousands to many millions of years, a time span that cannot be duplicated in the laboratory.

ORIGIN OF THE FIRST CELL

The last two sections described the enormous variety of organic molecules that are believed to have been produced during our planet's early evolution. Some of these molecules were probably similar or identical to those found in life today, while others were quite different.

Many survived for long times in sheltered niches, while others were soon broken down into simpler components by the same sources of energy that created them. The simple components were then reassembled into new, more complex molecules.

Initially, these chemical reactions occurred in a random, helter-skelter fashion and depended only on the prevailing physical conditions: the concentration of atomic and molecular raw materials, the available sources of energy for driving the reactions, and, possibly, the presence of catalysts such as clays, rocks, certain ions, and accidentally produced small protein-like molecules.

Eventually, an *organizing mechanism* or, as Manfred Eigen calls it, an "organizing principle" emerged among the reactions. This organizing mechanism included information and the means of translating that information into chemical function.

The emergence of this organizing mechanism was the beginning of a chemical evolution that, in the course of time, led to life. Today, such an organizing mechanism is present in the cells of all organisms, including those of our bodies. It consists of *genetic information* and *a biochemical machinery* that translates the information into *enzymes,* which, in turn, make the entire biochemical machinery run.

Before going on to the theory of how the organizing mechanism of life might have arisen, let us review how this mechanism works in today's cells. The components of the organizing mechanism are the following:

1. *DNA molecules* store *genetic information* by a code consisting of triplets of bases, called *codons*. The codons are arranged sequentially and partitioned into specific units, the *genes,* each of which codes for a particular protein molecule.

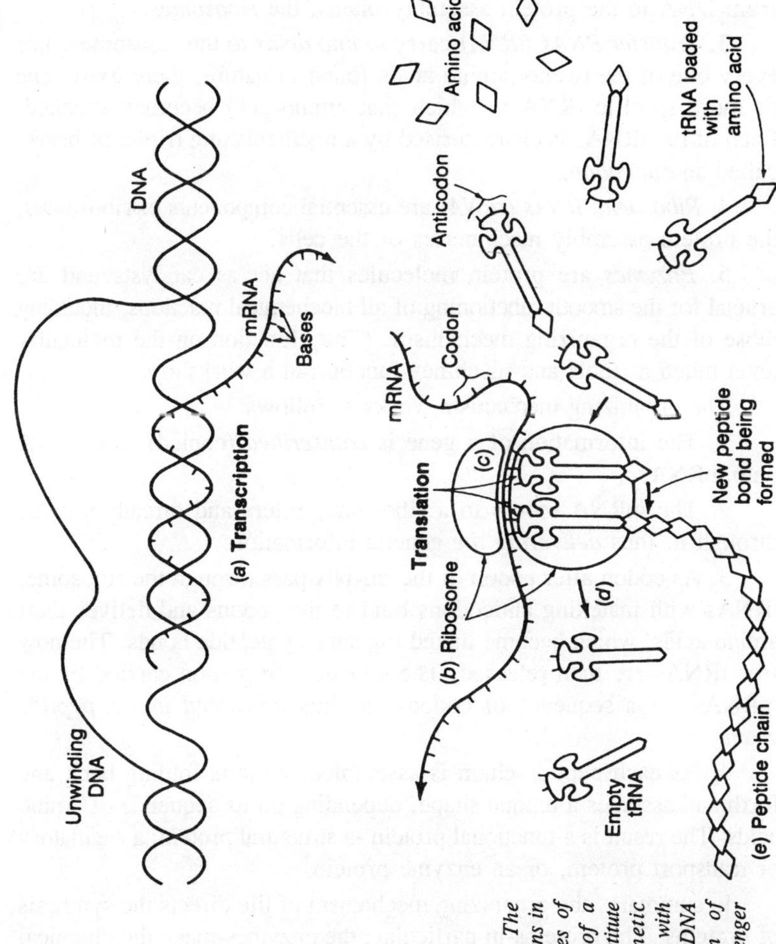

Figure 4.6 : Protein synthesis. The information for the synthesis of proteins in modern cells is carried by molecules of DNA. The information consists of a sequence of bases, three of which constitute a codon, the basic unit of genetic information. Protein synthesis begins with the unwinding of a double-stranded DNA molecule and the transcription of information onto a molecule of messenger RNA.

2. *Messenger RNAs (mR(VA)* carry the information of a given gene from DNA to the protein assembly plants, the *ribosomes.*

3. *Transfer RNAs (tRNA)* carry *amino acids* to the ribosomes. For every one of the twenty amino acids found in nature, there exists one or more specific tRNA to which that amino acid becomes attached. Each of the tRNAs is characterised by a distinguishing triplet of bases, called an *anticodon.*

4. *Ribosomal RNAs (rRNA)* are essential components of ribosomes, the protein assembly machineries of the cells.

5. *Enzymes* are protein molecules that act as catalysts and are crucial for the smooth functioning of all biochemical reactions, including those of the organizing mechanism. (They function on the molecular level much as tools and machines function in a workshop.)

The organizing mechanism works as follows:

1. The information of a gene is *transcribed* (copied) from DNA onto mRNA.

2. The mRNA moves to a ribosome, enters and threads its way through it, thus *delivering the* genetic information.

3. As codon after codon of the mRNA pass through the ribosome, tRNAs with matching anticodons bind to the codons and deliver their amino acids, which become linked together by peptide bonds. The now free tRNAs are then released. The genetic information carried by the mRNAs—as a sequence of codons—is thus *translated into a peptide chain.*

4. As each peptide chain is assembled, it starts folding back and forth and assumes a unique shape, depending on its sequence of amino acids. The result is a functional protein -a structural protein, a regulatory or transport protein, or an enzyme protein.

In summary, the organizing mechanism of life directs the synthesis of proteins. The proteins-in particular, the enzymes-make the chemical reactions in an organism work, including the chemical reactions of the organizing mechanism itself.

Origin of the Organizing Mechanism

The next question is, "How did the organizing mechanism of life have its start among the disorder and randomness of the chemical reactions on the early Earth?" Note that the organizing mechanism in contemporary life is based on the presence of both *informational* and *functional* components.

The informational components are DNA and RNA molecules. The

functional components are RNA molecules and proteins (enzymes). Interestingly, the RNAs have dual tasks: they carry information and they perform functions. The DNAs and proteins have single tasks each: the DNAs carry information and the proteins perform functions.

There are a number of reasons why in today's cells the RNAs, DNAs, and proteins have these particular features. RNA molecules are usually singlestranded, consisting of a linear sequence of monomers (the nucleotides, which contain the bases A, G, C, and U).

The sequential arrangement of monomers allows RNA molecules to carry information, similar to the way DNA molecules carry information. Furthermore, the single-strandedness allows RNAs to fold into three-dimensional shapes suitable for carrying out functions such as bringing amino acids to the ribosomes.

Some RNAs are also capable of catalyzing certain biochemical reactions and thus act like enzymes, as was recently discovered by a number of researchers. Such catalytic RNAs act either alone or as part of RNA protein complexes. Finally, short strands of RNA possess autocatalytic properties (see below). The three-dimensional folding of RNA has the additional advantage in that some of the shapes make the molecule resistant to destruction or alteration by hydrolysis and certain other chemical reactions.

In contrast, DNA molecules and proteins, the other components of the organizing mechanism in contemporary cells, do not have the RNAs' dual quality. The DNAs are generally double-stranded and, hence, are much less able than RNAs to fold into specific three-dimensional shapes that would allow them to carry out functions.

They are mainly used for storing and reproducing genetic information. We may assume, therefore, that DNA was initially not part of the organizing mechanism but was introduced later on. Proteins are made of peptide chains and resemble RNA in that they also fold into a great variety of shapes, which gives them their excellent functional qualities. However, peptides are not capable of making faithful replicas of themselves and, hence, are not suitable as information carriers.

It is unlikely that all of the components of life's organizing mechanism - RNAs, DNAs, and proteins (enzymes) -came into existence at once in the primordial soup. They must have evolved gradually from primitive and inefficient precursors that were barely distinguishable from accidentally formed molecules.

Here it is tempting to ask the old chicken and egg question, as biologists have since the 1930s: "Which came first, the informational or the functional components?" Experimental evidence gathered by Manfred Eigen and his coworkers, the American biochemist Leslie E. Orgel, and others suggests that neither came first. The informational and functional components of the organizing mechanism originated and evolved together. The start of this evolution is presumably to be sought among molecules of RNA because, of all the components of the organizing mechanism in contemporary cells, they alone carry information and perform functions.

RNA Quasi-Species

How did RNA molecules become the first component of life's organizing mechanism? A crucial laboratory experiment for finding an answer was carried out by Leslie Orgel. He demonstrated that short strands of RNA, such as U—U—U—...—U or C—C—C—...—C (poly-U and poly—C strands) are capable of making complementary copies of themselves in solutions containing activated nucleotides (that is, nucleotides with triphosphate chains).

For example, poly—C strands form poly-G strands (recall that G is the complementary base of C), and they do so with a fairly high degree of fidelity, meaning that very few errors (the substitution of A, U, or C for G, in this example) are introduced. An important feature of this experiment is that the RNA strands have autocatalytic qualities and their replication proceeds without enzymes.

This adds realism because, initially, enzymes would either not have been available in the primordial soup or, if present, would have been of poor and nonspecific quality. If zinc ions are added to the solution, much longer strands (up to 40 nucleotides in length) are copied and with greater fidelity than without those ions.

The zinc ions act as inorganic catalysts, which, very probably, were available in the soup. Interestingly, today's RNA polymerases (enzymes that catalyze the formation of RNA) all contain zinc ions, which led Eigen (1981, 101) to ask, "Has nature perhaps 'remembered' how replication started?"

Orgel's experiment demonstrated that strands of RNA are capable of selfreplication: They provide both the information and the function for making complementary copies of themselves. This result, as well as the dual roles RNA plays in contemporary cells, suggests that the first components in the evolution of life's organizing mechanism were indeed molecules of RNA or, at least, molecules similar to RNA.

We may imagine that life's organising mechanism got its start when short strands of RNA-perhaps only two or three nucleotides long

—were randomly assembled in the primordial soup. The original RNAs then began to replicate themselves by processes such as those in Orgel's experiment.

In the absence of well-developed enzymes, errors were commonly made during replication, despite the RNAs' short lengths. Hence, the replicated copies differed frequently in minor and major ways from the original ones with regard to sequences of bases and lengths. *Variation-the* first requirement of Darwinian evolution-was thus introduced right from the beginning into the pool of RNAs in the primordial soup.

Along with the creation of variation among the RNAs, *natural selection*—the second requirement of Darwinian evolution -got its start as well. The different kinds of RNA were not all equally successful in replicating themselves, avoiding errors during replication, and surviving.

For example, RNAs with an abundance of Cs and Gs probably tended to replicate themselves more faithfully than those with mostly Us and As; and RNAs with certain kinds of three-dimensional folding were more resistant to being broken up by water molecules and other

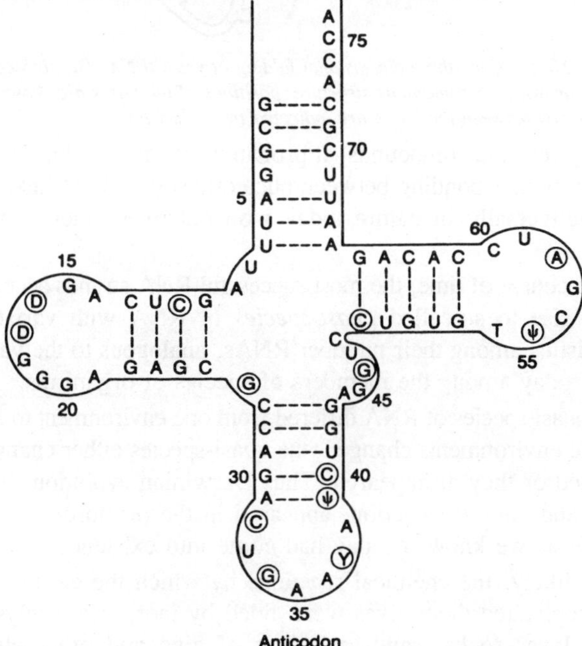

Figure 4.7 : Two-dimensional representation of the nucleotide sequence (76 bases) of yeast tRNA (coded for phenylalanine). It illustrates the clover leaf shape that is typical of all tRNAs.

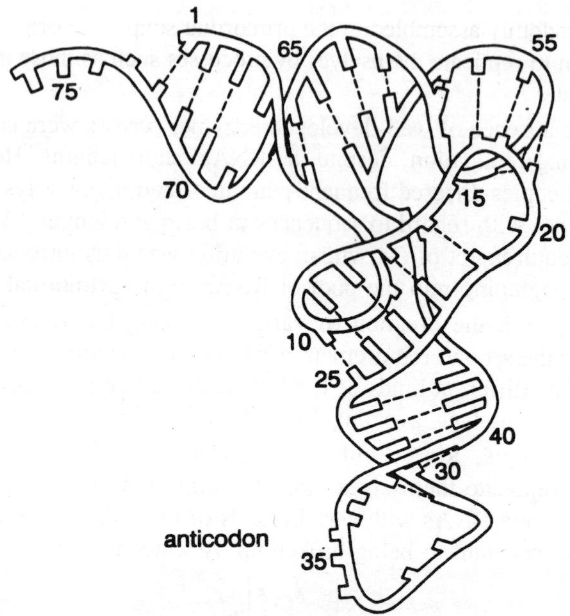

Figure 4.8 : Model of the three-dimensional folding of yeast tRNA. The ribosephosphate backbone is drawn as a continuous structure; positions of the bases are shown as bars; and the H-bonds between the bases are indicated by dashed lines.

chemically reactive compounds. It probably was also at this early stage that the particular bonding between nucleotides (the 3'-5' linkage) that we find universally in nature today won out over other, competing bonds.

In the course of time, the most successful RNA strands accumulated and gave rise to so-called *quasi-species* of RNA with variations in characteristics among their member RNAs, analogous to the variations observed today among the members of species of organisms.

The quasi-species of RNA differed from one environment to another; and, as the environments changed, the quasi-species either changed also and adapted or they disappeared. Thus Darwinian evolution, based on variation and natural selection, appeared in the primordial soup long before life as we know it today had come into existence.

Very likely, the chemical reactions by which the earliest strands of RNA replicated themselves were aided by inorganic catalysts such as clays, lava, rocks, sand, and ions of zinc and other elements. Accidentally produced protein structures such as Fox's proteinoid microspheres might have placed a role as well.

In the absence of specific enzymes that catalyzed replication, Eigen estimates that the early RNAs could at most have been about 50 to 100 nucleotides long, which is comparable to the lengths of today's tRNAs. Had they been much longer, the number of errors introduced during replication would have been so great that within a few generations their particular base sequences would have become changed be yond the range of viability.

RNA Hypercycles

As long as the strands of RNA consisted of no more than 100 nucleotides, they could do little else than help in their own replication. They were not long enough to store information for the assembly of enzyme catalysts. However, enzymes were needed to allow the organizing mechanism to progress further.

They were needed for the faithful replication of longer RNAs, and the longer RNAs were needed to store the information for the assembly of the enzymes. Furthermore, specialised functional RNAs were needed, analogous to today's tRNAs (and possibly rRNAs), to help in the translation of the information into enzyme structure.

According to theoretical studies by Eigen and his coworkers, the evolution toward longer strands of RNA and enzymes required the development of cooperative couplings between different RNA quasi-species. A simple example of such couplings involves three quasispecies. The RNAs of one of the quasi-species carry information for the assembly of enzymes (albeit, very primitive ones) and the RNAs of the other two quasi-species function as tRNAs.

With the aid of the tRNAs, the information of the first quasi-species is translated into primitive enzymes. The enzymes, in turn, aid in the replication of the RNAs of all three quasi-species. Such cooperative couplings are called *hypercycles*.

In reality, hypercycles in the primordial soup probably were never as simple as the one just described. We may assume that many more quasi-species of RNAs and kinds of enzymes than those indicated in the example were required, even for inefficient hypercycles. Hypercycles probably evolved from single RNA quasispecies that consisted of large numbers of RNAs with the usual variations with regard to length and base sequence.

At first the participating tRNAs, helping in the assembly of peptides according to information carried by the longer RNAs, were quite inefficient. Likewise, the peptide chains, which functioned as enzymes, were quite inefficient also. They were probably not much more effective

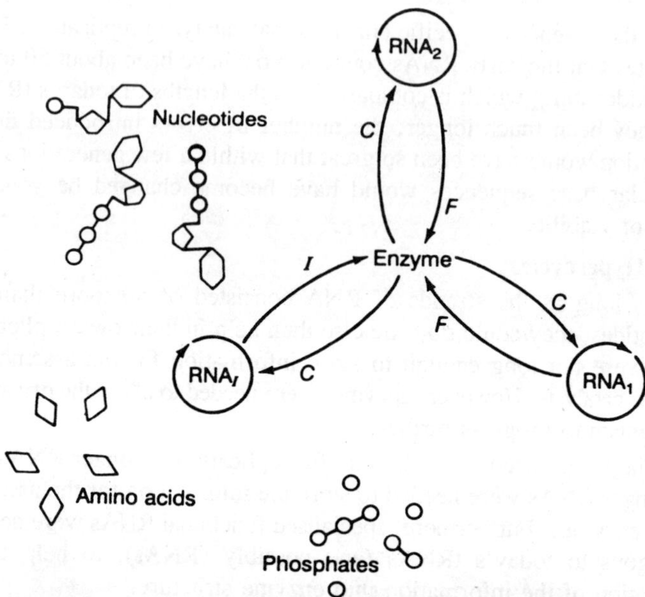

Figure 4.9 : Diagrammatic representation of a hypercycle, consisting of three distinct RNA quasi-species (labeled RNA,, RNAI, and RNA2) and enzymes.

in this task than many of the accidentally produced peptide chains and other inorganic catalysts. In the course of time, however, the lengths of some of the participating RNAs increased, and their information content-coding for the assembly of enzymes -grew more distinct.

Other RNAs remained short (50 to 100 nucleotides) and took on the role of today's tRNAs. The enzymes thus produced became more specific and efficient in their tasks. Slowly the original RNA quasi-species evolved into a number of distinct RNA quasi-species, with each quasi-species specializing in a particular task: to store information for the assembly of specific enzymes or to function as tRNAs.

Thus the different RNA quasi-species, along with their enzymes, became firmly coupled and dependent on each other for survival. They evolved into hypercycles.

The longest RNAs of hypercyclic quasi-species were conceivably up to several thousand nucleotide units in length, as suggested by the gene lengths of certain contemporary viruses. That was roughly a ten-to-fiftyfold increase over the lengths of the RNA strands of the earlier, independent RNA quasi-species.

This increase was possible because the RNAs of the hypercycles were replicated with the aid of specific enzymes, which greatly improved copying fidelity and efficiency. We may assume that at this stage of prebiotic evolution the modern triplet code for storing genetic information came into existence as well. The code was needed to translate accurately the information carried by RNAs into enzyme structure.

However, at present it is not clear whether the triplet code started out that way or was preceded by singlet or doublet codes. Singlet or doublet codes would have made the initial stages of the development of the organizing mechanism easier, but it is difficult to see how the switch from a singlet or a doublet code to a triplet code could have come about, because such a switch would have invalidated much of the previously accumulated information.

Nevertheless, some researchers believe that the first code was singlet or doublet and that somehow, at a later stage during the origin of life, it became converted into the triplet code that is in universal use today.

Prebiotic evolution progressed from independent RNA quasi-species to hypercycles not only because the latter produced their own enzymes. It also progressed in this direction because, through the enzymes, the participating RNA quasi-species became dependent on each other and were forced to cooperate for the common good.

They could not afford to outcompete and to eliminate each other. To survive, they had to cooperate. This cooperation greatly increased their reproduction rate and made them the dominant quasi species in the soup.

The advantages of hypercyclic cooperation may be illustrated by an example from economics. Let us imagine a primitive human society in which every family is self-sufficient and competes with other families for available raw materials. Each family grows its own food, makes its own tools, builds its own shelters, and defends itself against competitors.

Such an economy may be compared with the earlier, noncooperating RNA quasi-species, which competed with each other for the available raw materials. A much more efficient and productive economy results when the families begin to join into cooperative units, with each family specializing in specific tasks.

Some families are farmers and grow food. Others are blacksmiths, butchers, bakers, or carpenters and specialize in the making of tools, slaughtering, baking, and building. Such a cooperative economy will easily outproduce and outcompete the simpler one in which each family fends for itself. The same was true of the hypercycles of coupled and cooperating RNA quasi-species, in which the different kinds of participating RNAs had specialised informational and functional tasks and in which they cooperated for the common good.

Another example of hypercycles is our present-day ecosystem, with its many interdependent and cooperative linkages among animals, plants, and microorganisms. Without the support from microorganisms, animals and plants could not exist. Likewise, many microorganisms depend on plants and animals for their nutrients.

Furthermore, all animals depend (directly or indirectly) on plants for food and oxygen. In analogy, we may regard the RNA quasi-species that were linked into cooperating hypercycles in the primordial soup as our planet's first ecosystem.

Protocells

The arrival of hypercycles of coupled and cooperating RNA quasi-species constituted an important step forward in the development of life's organizing *mechanism*. However, there still remained a serious obstacle. There still existed no decisive feedback from the enzymes to the information content of the RNAs.

Natural selection did not favour RNAs because they carried information for superior enzymes; it favoured RNAs that were the most stable and reproduced most rapidly under the prevailing conditions, regardless of whether they also carried useful genetic information or not. Even RNAs that carried no information at all and performed no functional tasks benefited from the enzymes produced by other RNAs.

This feedback problem resulted from the fact that up to now all chemical reactions had taken place in the open environment of the primordial soup. As long as this remained the case, the obstacle could not be overcome. Only by enclosing the hypercycles with membranes and forming protocells did evolution progress further.

Competition could then occur between the protocells, which forced natural selection to focus on the fitness of the protocells' entire enzymedriven biochemical machineries. Cells that survived possessed, by definition, the most advantageous enzymes, and (because the cells survived) their genes were passed on to the next generation.

Thus, the feedback from enzyme to gene—or from phenotype to genotype—that is characteristic of all contemporary life became established. Besides solving the feedback problem, membrane enclosure had other advantages. It kept the enzymes produced by the hypercycles close to the RNAs and prevented their diffusion into the environment.

It also created the possibility of concentrating biochemical raw materials -such as amino acids, nucleotides, and phosphates -by the development of selective transport mechanisms into the cells' interior. At present no one knows just how the evolution from hypercycles to protocells came about.

It was a step that probably required the parallel evolution of numerous major and interconnected biochemical capabilities: the evolution of suitable membranes; selective transport of the right kinds of raw materials through the membranes into the cell interiors; elimination of waste products; extraction of energy from energy-rich molecules; replication of genetic information with high fidelity; and efficient translation of the genetic information into enzymes and other kinds of proteins.

Though we do not know how these capabilities arose, it is clear that to function reasonably efficiently protocells required many more enzymes than were produced by the earlier hypercycles. Furthermore, these enzymes needed to be constructed with an ever higher specificity for catalyzing particular chemical reactions.

Eventually, the information required for assembling the enzymes became so great that it could not have been stored and passed on from generation to generation if strands of RNA had remained the only information carriers. Too many errors would have been made during the replication of RNA, despite the assistance by enzymes, and any one of the errors might have been fatal to the cells. The only way out was the development of an error suppression mechanism.

The error suppression mechanism that evolved depended on the introduction of a double-stranded information carrier. That information carrier was DNA. As noted before, the two strands of DNA are complementary-the base adenine (A) of one strand is always bonded to thymine (T) on the other strand, and cytosine (C) is always bonded to guanine (G). Hence, the two strands carry identical information.

The presence of the same information twice permitted the development of a "proofreading" mechanism during replication. While a copy is made of one of the strands of DNA, it is checked by proofreading enzymes against the other, complementary strand.

Any errors detected are corrected. This *error suppression* mechanism operates in all contemporary cells. It must also have evolved, at least in primitive form, in the protocells. It allowed their DNAs to store genetic information that extended over hundreds of thousands to millions of nucleotides, which probably was enough for the protocells' genetic requirements.

The DNAs were replicated sufficiently faithfully that errors -that is, mutations—did not, in general, accumulate to harmful levels. The occasional errors that did occur contributed to the genetic variations of the protocell species and, hence, to their evolution.

Very likely, species of protocells arose separately on numerous occasions in the primordial soup. Nevertheless, the universality today of the structures of proteins and nucleic acids, the triplet code, the use of ATP, and the workings of the organizing mechanism suggest that competition and natural selection eliminated all but one of the early species of cells. The surviving species became the ancestors of all subsequent terrestrial life -from bacteria to protists, fungi, plants, and animals.

As discussed in the following chapter, there are good reasons to believe that the earliest cells were prokaryotes and that they reproduced by binary cell division, similar to the reproduction of contemporary bacteria. Furthermore, they probably derived their energy by fermenting energy-rich molecules, such as sugars, that they found in the soup.

No one knows how much time elapsed before chemical evolution made the transition to the first species of cells. It may have been a million years or less. Or it may have been many hundreds of millions of years.

All we know, from the fossil record, is that by about 3.5 billion years ago single-celled life inhabited shallow areas of the sea. There are also indications that these cells were capable of photosynthesis and, hence, had already evolved considerably beyond the state of the earliest fermenters. For lack of more specific information, let us assume then, somewhat arbitrarily, that protocells arose approximately 4 billion years ago.

This is a round figure that will suffice for discussions in the next chapter, even though it may be off by a few hundred million years one way or the other. With the arrival of protocells in the primordial soup, the fundamental components of life as we know them today had come into existence, at least in rudimentary form: the organizing mechanism, consisting of DNAs, RNAs, and enzymes; the triplet code for writing genetic information; the means of selective absorption of raw materials and of expulsion of waste products across the cells enclosing membranes; the extraction of energy from energy-rich molecules; and the capacity for cell reproduction and passing genetic information on to the next generation.

The driving mechanism in the development of the protocells was Darwinian evolution. According to the theory presented here, this evolution began with the self-replication of short strands of RNA among the otherwise random chemical reactions in the primordial soup and progressed step-by-step to RNA quasi-species, hypercycles, and membrane enclosure of some of the hypercycles.

Remember that the theory is based on many assumptions and contains many gaps. Some of the assumptions may turn out to be wrong. In fact, researchers in a number of laboratories today are developing and testing alternative theories, some of which rely on RNA and RNA like molecules as the first replicator and other that don't.

Clearly, investigation of the origin of life is at present characterised by great uncertainties and enormous intellectual challenges. Nevertheless it is an exceptionally exciting field of scientific research, dealing as it does with one of nature's most fundamental secrets.

5

Evolution of Early Life

With the origin of life on Earth approximately 4 billion years ago, a new era began in the evolution of matter. The physical laws were no longer the only ones determining the course of events. The laws of biology—based on the random production of variation and on natural selection—exerted their influence now as well.

Over the eons, early life became more complex and diversified. Competition in the struggle for survival intensified and new ways of utilizing available raw materials and energy evolved. Gradually, life spread to all parts of our planet's surface, interacting with the physical environment and changing its composition, structure, and appearance. A worldwide ecosystem became established and the greening of the Solar System's third planet got under way.

When we keep in mind the humble beginnings of life, these far-reaching effects are both surprising and impressive. The protocells were primitive, unicellular prokaryotes that lived off organic molecules they found in the primordial soup.

They metabolised their food by fermentation, much as some bacteria and yeast cells do today, and they reproduced by simple binary cell division. Photosynthesis, the direct use of the energy of sunlight to manufacture organic molecules, had not yet evolved. Nor had respiration, the major metabolic pathway of the cells of all animals, plants, and other complex multicelled organisms.

In fact, as discussed in chapter elsewhere in this chapter, the primary prerequisite for our modern kind of respiration, molecular oxygen (O_2), did not exist initially in the Earth's atmosphere. Had it been present, the early cells would have been killed by its high chemical reactivity. Finally, the eukaryotic cell and reproduction by sexual mating had not yet developed either.

A SURVEY OF THE EVOLUTION OF EARLY LIFE

The evolution of photosynthesis, respiration, eukaryotic cell structure, and eukaryotic sexual reproduction was slow in coming. The first step in this development was taken when the early fermenting prokaryotes began to spread across the primordial soup and multiplied to the point where the organic food supply—which initially was produced by abiotic, random chemical reactions—ran short. The competition for new sources of raw materials and energy intensified.

Photosynthesis

Some bacteria adapted by evolving ways to trap the energy of sunlight directly and use it to synthesize their own food from inorganic raw materials. They obtained the necessary carbon and oxygen from carbon dioxide (CO_2) and the hydrogen from such hydrogen-rich molecules as hydrogen sulfide (H_2S) and molecular hydrogen (H_2). *Photosynthesis* thus evolved.

However, unlike the photosynthesis carried out by modern blue-green bacteria (more about them shortly), algae, and green plants, the photosynthesis of those ancient bacteria did not release 02. It was an *anaerobic* kind of photo synthesis, probably similar to the one still found today in the purple and green sulfur bacteria and the purple nonsulfur bacteria. There are indications in ancient rock records that anaerobic photosynthetic bacteria lived in the primordial soup 3.2 to 3.5 billion years ago.

In contrast to the fermenting heterotrophs, the photosynthetic bacteria were *autotrophs,* meaning they could "nourish themselves" starting with inorganic raw materials. They did not depend on already synthesized organic nutrients. In the course of time, as the early photosynthesizers became more efficient, they produced more and more of the organic matter in the soup. Eventually, they became the primary producers in the food chain of ancient life.

The development of photosynthesis required the simultaneous invention of two new biochemical *tools-chlorophyll* (or some other molecules with similar light-absorbing properties) and the *electron-transport chain.* Just as it is in contemporary photosynthesizers, chlorophyll was the actual light-gathering molecule by which ancient photosynthetic bacteria trapped the energy of photons and boosted electrons to high-energy states.

The electron-transport chain was the tool by which the energetic electrons were returned to the unexcited, low-energy state in a sequence of small steps. This happened via a series of carrier molecules which, in bucket brigade fashion, passed the energetic electrons from one to the other with an accompanying release of energy.

Some of the released energy was stored in molecules of ATP and became available for useful work, such as the synthesis of organic molecules. The rest of the released energy was lost as heat.

As discussed below, both chlorophyll and the electron-transport chain were crucial for the development of still more advanced metabolic pathways during the succeeding evolution of early life (also see sections, The Evolution of Photosynthesis and The Makingof the Eukaryotic Cell).

Despite their independence from the vanishing supply of organic nutrients in the soup, anaerobic photosynthetic bacteria still suffered from one serious shortcoming. They depended on molecules like H_2S and H_2 for their requisite supply of hydrogen atoms.

Although H_2S and H_2 were available, as they still are today, in the swamps and in the vicinity of volcanoes and thermal pools of the early Earth, they were not nearly as universally abundant as was another hydrogen-rich molecule-water (H_2O).

The problem with H_2O was that it is a particularly stable molecule and does not give up its hydrogen atoms as readily as do H_2S and H_2. A new and more powerful kind of photosynthetic apparatus was required to split H_2O into its atomic constituents. In the course of time, some bacteria evolved such an apparatus.

Like their anaerobic photosynthetic predecessors, they used the hydrogen atoms in the synthesis of carbohydrates and other organic products. They released the oxygen, derived from the water, as O_2 into the atmosphere, where it gradually accumulated. This kind of metabolism is *aerobic photosynthesis*.

Aerobically photosynthesizing bacteria began to proliferate about 2.4 billion years ago and subsequently expanded into virtually all ecological niches containing water and sunlight, replacing the anaerobic photosynthesizers as the primary food producers on our planet.

Their modern descendents-the blue-green bacteria, algae, and green plants—have maintained this position to this day.

Respiration

At the time when photosynthesis was emerging, some bacteria evolved in a rather different direction. They acquired the ability to derive energy from reactions between various kinds of organic compounds and such inorganic molecules as sulfate (SO_4), nitrate (NO3), and carbonate (CO_3).

This new kind of energy generation is called *anaerobic respiration,* where "respiration" refers to the fact that inorganic molecules participated in releasing the energy stored in the organic compounds, and "anaerobic" means that no O_2 was involved.

The metabolic pathways of the anaerobic respirers differed greatly from one bacterial species to the next, depending on the organic and inorganic raw materials used by them. Some of these unusual types of anaerobically respiring bacteria have survived with few changes to this day as, for example, species of *Desulfovibrio* (sulfate respirers), *Pseudomonas* (nitrate respirers), and methane-producing bacteria (carbonate respirers).

The reactions of anaerobic respiratory metabolisms, which combined organic molecules with oxygen-rich inorganic compounds, released so much energy that they were potentially harmful. Cells overcame this danger by adopting the electron-transport chain, which had already evolved earlier in the photosynthetic bacteria.

Through the electron-transport chain, bacteria were able to control the chemical reactions of anaerobic respiration and let energy come off in a series of small steps, rather than all at once. As in photosynthesis, some of the released energy was stored in molecules of ATP and became available for useful work.

The application of the electron-transport chain in the respiratory pathways of anaerobic bacteria was a crucial preparatory step for the next major evolutionary step—the development of *aerobic respiration.* Aerobic respiration also required organic foodstuffs, but instead of depending on such inorganic molecules as SO_4, NO_3, or CO_3, it utilised O_2, which by then had become abundant in the atmosphere and oceans of the Earth.

The oxygen molecule was split into two oxygen atoms, which were then allowed to react with hydrogen atoms to form water. The advantage of this reaction was that it released far more energy than did anaerobic respiration or any of the other metabolic pathways developed earlier. As in anaerobic respiration, the reaction between hydrogen and oxygen atoms was carried out safely with the aid of the electron—transport chain. The development of aerobic respiration was completed between about 2.0 and 1.5 billion years ago.

Eukaryotic Cells

Aerobic respiration, with its ability to liberate large amounts of energy through reactions between hydrogen and oxygen, became the basis for further evolutionary progress that eventually led to modern *eukaryotic cells.* Eukaryotic cells distinguished themselves from their prokaryotic ancestors by having *nuclei* that enclosed the genetic material.

That material was distributed over several pieces of DNA, namely the *chromosomes.* All of the eukaryotes metabolic activities involving 02 were carried out within special organelles—mitochondria (respiration) and chloroplasts (photosynthesis). In addition, eukaryotic cells evolved other organelles that were responsible for synthesizing proteins, packaging and transporting proteins, and many other tasks, as discussed in the Introduction to Part Two.

With the development of eukaryotic cells, the binary cell division of prokaryotes evolved into a more complex process called *mitosis.* Mitosis ensured that the replicated chromosomes were properly divided into two equal sets and passed on to the offspring cells. Eventually, mitosis evolved into yet another kind of cell division-meiosis. Meiosis produced reproductive cells and was the basis for *eukaryotic sexual reproduction.*

The arrival of eukaryotic cells was a pivotal event in the evolution of life. Early prokaryotic organisms were microscopic and unicellular, and with few exceptions they have remained so to this day. Eukaryotic organisms, in contrast, had the potential to evolve into more complex multicellular and macroscopic forms, and within a few hundred million years of their arrival they realised this potential. By approximately 700 million years ago, the first jellyfish, worms, and other soft-bodied animals began to populate the waters of the Earth.

If life came into being approximately 4 billion years ago, the evolution from the first fermenting prokaryotes to complex multicelled life took roughly 3.3 billion years: about 500 million years for the development of anaerobic photosynthesis, another 1.1 billion years until 02-releasing photosynthetic bacteria began to proliferate, still another 900 million years until aerobic respiration and the eukaryotic cell had evolved, and, finally, 800 million years more to the arrival of the first complex multicellular organisms.

Another way to think about this evolution is to imagine that the age of the Earth is compressed into one year and to assume that our planet

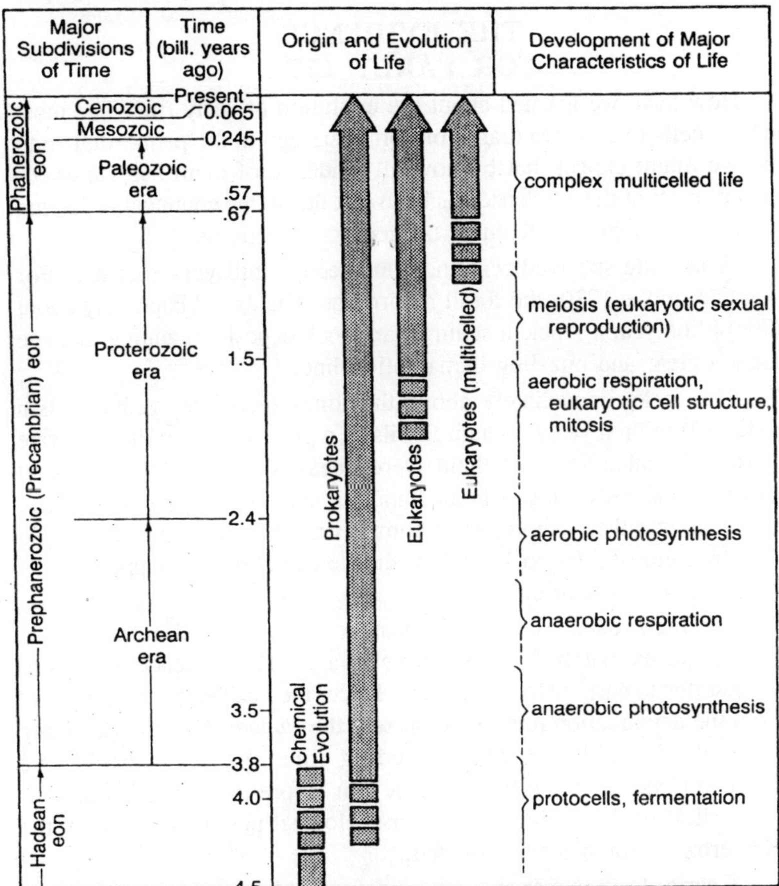

Figure 5.1 : Evolution of prokaryotic and eukaryotic life.

was formed on January 1 (when, according to this time scale, the Universe was already three years old). During much of January, the Earth acquired its early anaerobic atmosphere and juvenile ocean.

The first fermenting prokaryotes came into existence in early February. Photosynthesis was invented in March. Free oxygen started to become abundant by mid July. Aerobic respiration and the eukaryotic cell structure developed during August. Sexual reproduction appeared some time in September.

It was late October before complex multicellular organisms appeared in the sea. And the evolution of plants and animals into the forms that we see today took place during the final eight to nine weeks of this imaginary year.

THE EVIDENCE FOR EARLY LIFE

How have we learned about the evolution of early life? The most ancient cells lived more than 3 billion years ago in the primordial soup and we might expect that by now all evidence of them has vanished. Indeed, most of that evidence has been lost due to the continual reshaping of the Earth's surface features by geological activity.

What little survived remained undetected until very recently. For example, in the 1930s the fossil record could be traced back only about 570 million years. Ancient sedimentary rock deposits spanning that age show a clear and puzzling demarcation line.

The rocks immediately above that line—therefore, younger than about 570 million years-contain fossils of a great variety of invertebrate marine animals. Some of them were encased in shells or possessed skeletons; others were soft-bodied and without any hard parts. Clearly, the ocean was then teeming with complex multicelled and macroscopic life. In contrast, the rocks below the demarcation line appear to be devoid of any fossil remains.

Did life suddenly come into existence, fully shaped in multicelled forms, approximately 570 million years ago? The evidence seemed to suggest this to early paleontologists. Hence, they called the time interval from the demarcation line to the present the *Phanerozoic eon,* meaning "eon of visible life." The preceding time span they called the *Precambrian eon,* because it was the eon before the Cambrian period, the earliest of the traditional eleven geological periods into which the Phanerozoic Eon has been divided.

Charles Darwin was aware of and puzzled by the abrupt appearance of fossils of complex, macroscopic life at the beginning of the Phanerozoic eon, and in *On the Origin of Species* he wrote:

> To the question why we do not find rich fossiliferous deposits belonging to ... periods prior to the Cambrian system, I can give no satisfactory answer.... The case at present must remain inexplicable; and may be truly urged as a valid argument against the views here entertained.

Today we know that the Precambrian rocks are not devoid of fossil evidence, but much of that evidence stems from microscopic, unicellular organisms and so is not easily seen. The specialised laboratory techniques required to detect it were developed only during the 1950s. Since then microfossils have been found in dozens of Precambrian rock outcrops on all continents, from the Grand Canyon and the Rocky Mountains to many other sites in South Africa, Australia, Siberia, and elsewhere. In addition to microfossils, evidence for the existence and evolution of Precambrian life comes from the effects it had on the physical environment.

For example, some ancient rock deposits contain rusted iron and other metals, which implies that molecular oxygen was present, presumably released by early photosynthetic bacteria. Other ancient rocks contain unique isotopic abundances of carbon and other elements that can be explained only as being due to biological activity.

Yet a third line of evidence of the evolution of Precambrian single-celled life has been found in our own cells and those of other eukaryotes and prokaryotes, through a new understanding of their metabolic pathways and other biochemical processes. For example, when we put forth a sudden physical effort, like running a 100-meter dash or rapidly climbing a flight of stairs, our cells temporarily run out of 02 (that is, we run out of breath) and are forced to halt or slow down their aerobic respiration.

They then generate ATP (energy) anaerobically, much as fermenting bacteria do. Our cells can obtain energy in this manner for only short durations, but while they do they are reverting to the most ancient of all metabolisms-fermentation-from which all others have evolved.

None of the three sources of evidence -Precambrian microfossils, the ancient inorganic rock record, and biochemical processes in contemporary cells - would alone suffice to deduce the early history of life with any degree of confidence.

Together, however, they point rather compellingly to the step-bystep evolutionary progression outlined in the previous section, from the first fermenting prokaryotes to the development of photosynthesis, respiration, eukaryotic cell structure, eukaryotic sexual reproduction, and, finally, the arrival of complex multicellular organisms. The remainder of this chapter will discuss these sources of evidence in more detail.

THE GEOLOGIC EVIDENCE

All continents include areas of low-lying rock outcrops that were laid down as sediments in the oceans during Precambrian times. Organisms that lived in the ancient oceans frequently became embedded in the sediments. Additional deposits in later eras often buried the older sediments to depths of thousands of meters, creating enormous pressures and temperatures and fusing them into compact, crystallised rocks (by a process called *metamorphism*).

Simultaneously, whatever organismic remains the buried layers contained got crushed and destroyed. However, in some locations the burial of the ancient rocks never amounted to very much. Consequently, they did not experience extreme heating and compression, and some of the organisms trapped in them were preserved as recognizable fossils.

Microfossils

The most ancient fossils of early life discovered to date have been found in sedimentary rocks at North Pole in western Australia, with an estimated age of 3.5 billion years (as deduced from the decay of long-lived radioactive isotopes). These fossils consist of mound-shaped structures resembling modern stromatolites, simple microspheroids, and complex filamentous microfossils.

However, as the discoverers of these fossils (most of them are from the University of Western Australia and the Australian Bureau of Mineral Resources, Geology, and Geophysics) point out, they cannot be absolutely certain that the mound-shaped structures and microspheroids are of biological origin. And the rocks containing the filamentous microfossils appear to have been contaminated by carbonaceous matter some time after sedimentation took place, a process that may have introduced the microfossils.

But most experts are beginning to agree that these objections are not valid and that the North Pole fossils are indeed 3.5 billion years old. Fossils with a reported age only slightly less than that of the North Pole specimens come from the Onverwacht and Fig Tree cherts of South Africa. The ages of these rocks have been estimated at 3.4 billion years for the Onverwacht formation and 3.2 billion years for the Fig Tree formation.

The Onverwacht microfossils, which are spheroidal or filamentous in shape, range in size from approximately 4 to 20 μm, which is similar to the sizes of modern bacteria. The fossils from the Fig Tree formation are spheroidal or rod-shaped. The spheroids have sizes like those from the Onverwacht site; the rods are somewhat smaller, with diameters of about 0.3 μm and lengths from 0.5 to 0.7 μm.

A few of the Fig Tree fossils show double-layered membranes comparable in thickness to those of modern bacteria. The spheroidal fossils bear some resemblance to modern blue-green bacteria and may be the remains of aerobic photosynthesizers or of their precursors.

Despite these microfossils' similarity to contemporary bacteria, both in size and shape, a word of caution is in order here also. All of them show a very simple morphology and, in the words of William

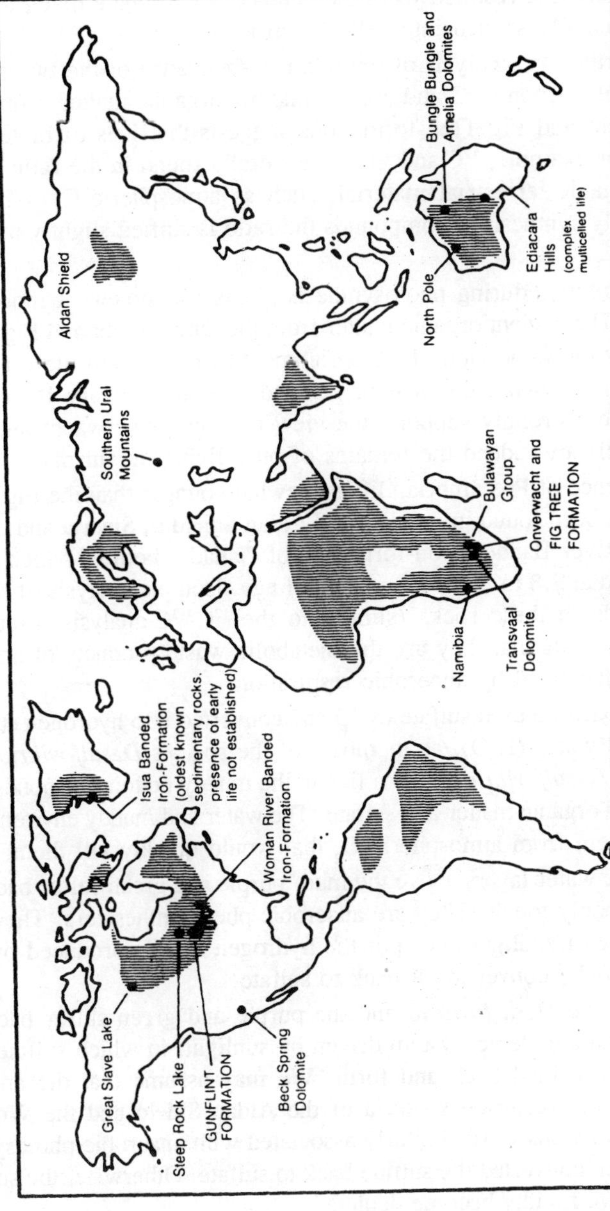

Figure 5.2: Selected sites of ancient rock formations that contain or are suspected to contain fossilized remains of early life. All continents include areas of low-lying rock outcrops that were laid down as sediments in the oceans billions of years ago and later became fused into rocks by additional deposits above them. These areas are referred to as shields and are indicated in the above map by shading.

Schopf, "do not exhibit a degree of structural complexity that would make their biological interpretation wholly convincing." They may be pseudofossils that resulted from the deposit of carbonate material that was accidentally shaped into cell-like structures.

Nevertheless, analysis of the relative abundance of the two stable isotopes of carbon, ^{12}C and ^{13}C, found in organic matter from the Onverwacht and Fig Tree formations suggests that it is of biological origin. For example, ^{12}C and ^{13}C are generally found in the ratio 99 to 1 in inorganic terrestrial material, such as atmospheric CO_2, but in biologically synthesized compounds the ratio is shifted slightly toward ^{12}C.

Apparently, during photosynthesis plants fix somewhat more ^{12}C than ^{13}C. The ancient organic matter from the Onverwacht and Fig Tree sites show the same kind of ^{12}C enrichment as is found in more recent deposits of known biological origin and in contemporary biological matter. This strongly supports the view that the Onverwacht and Fig Tree fossils are indeed the remains of once-living organisms.

Evidence of Precambrian life somewhat younger than the Fig Tree fossils has been found in rocks of the Aldan Shield of Siberia and of the Woman River Banded Iron-formation of Canada, both of which have been dated at 2.8 to 3.1 billion years of age. Isotopic analysis of sulfur compounds in these rocks (similar to the $^{12}C/^{13}C$ analysis discussed above) suggests that they are the metabolic waste products of ancient bacteria that lived by anaerobic respiration.

The bacteria used sulfate (SO_4) and converted it to hydrogen sulfide (H_2S) and water (H_2O), as members of the species *Desulfovibrio* still do today. *Desulfovibrio* bacteria live in the muddy bottoms of ponds that are rich in organic matter and sulfate. The water and muddy environment protect them from atmospheric O_2 that would quickly kill them.

In the water layers above the mud, purple and green sulfur bacteria are commonly found. They are anaerobic photosynthesizers: They use the energy of sunlight and split the hydrogen sulfide produced by the *Desulfovibrio,* converting it back to sulfate.

Thus the *Desu fovibrio* and the purple and green sulfur bacteria constitute a simple ecosystem driven by sunlight, in which sulfate and sulfide are cycled back and forth. We may assume that the ancient anaerobically respiring bacteria of the Aldan Shield and the Woman River formations were similarly associated with anaerobic photosynthesizers that converted the sulfide back to sulfate. Otherwise, the sulfate would have rapidly become depleted.

However, be cautious, as with the interpretation of the North Pole, Onver wacht, and Fig Tree fossils, and keep in mind that the evidence for the existence of ancient sulfate respiring and anaerobic photosynthetic bacteria is not absolute. It is an indirect kind of evidence and requires further research to resolve whether those ancient sulfur compounds are indeed metabolic waste products or have some unknown abiotic origin.

Stromatolites

In contrast to the record of Precambrian life discussed so far, much of the evidence found in younger rocks clearly shows its biological origins. Good examples are the unusual dome-shaped fossil structures that were discovered around the turn of the century by the American paleontologist Charles D. Walcott in the 2-billion-year-old Gunflint Iron-formations along the western shores of Lake Superior in Canada. These structures look like tall stacks of pancakes and reminded Walcott of modern stromatolites.

Stromatolites are matlike, sedimentary deposits in the form of domes or pillars, which today are the result of the life cycle of certain kinds of blue-green bacteria. The bacteria either trap sediment grains in a jellylike material they secrete or their metabolic activity causes sediments to precipitate. Because blue-green bacteria are aerobic photosynthesizers, Walcott suggested that the

Gunflint fossil structures are also the product of aerobic photosynthetic bacteria. At first his suggestion was received with skepticism, largely because at the time there existed little other evidence that life dates back billions of years. Then in the 1950s, American paleobiologists Stanley A.

Tyler and Elso Barghoorn, discovered microscopic fossils in the Gunflint stromatolites that were undoubtedly the remains of living organisms and nearly identical to contemporary blue-green bacteria, and the doubt about the origin of the fossil stromatolites vanished. Today it is commonly accepted that they had indeed been built up in shallow water by aerobic photosynthetic bacteria, like their modern counterparts.

During the last fifteen years, fossilized stromatolites have been found in dozens of locations all over the globe. Many are of the age of those of the Gunflint formations or younger, but a few are older, such as those of the Bulawayan Group in Rhodesia (about 3.1 billion years in age), Steep Rock Lake in Canada (2.6 billion years), the Transvaal Dolomite of South Africa (2.3 billion years), and near the Great Slave Lake, Canada (2.0 billion years).

These findings suggest that aerobic photosynthesis may have evolved very early, though it probably did not become widespread until about 2.4 to 2.0 billion years ago.

Two Precambrian Eras

The proliferation of aérobic photosynthetic bacteria about 2.4 billion years ago and the accompanying release of O_2 into the atmosphere, as manifested by ancient fossil and inorganic rock records, offers a convenient way of subdividing the long Precambrian eon into two eras: the *Archean* era, spanning the time from the oldest known sedimentary rocks, with an age of about 3.8 billion years, to 2.4 billion years ago; and the subsequent *Proterozoic* era, extending from 2.4 billion years ago to the beginning of the Phanerozoic eon, which is traditionally recognised as about 570 million years ago. *(Proterozoic is* derived from the Greek and means "before visible [macroscopic] life.")

In summary, the Archean era is the oldest time period for which there is evidence of the existence of life on Earth. Relatively few fossils have survived from this era, and the few that have are poorly preserved and show only very simple cell structures.

Therefore, some doubts remain whether they are, in fact, remains of organisms or have some unexplained abiotic origin. In contrast, many of the microfossils of the Proterozoic era are well preserved and leave no question as to their biological origin.

They attest to the progressive diversification, proliferation, and increase in structural complexity of early life. In particular, they indicate that the most dominant organisms of the Proterozoic era (at least for most of its duration) were the blue-green bacteria. Hence, this era may justly be called the "age of the blue-green bacteria."

Inorganic Rock Record

The arrival of aerobic photosynthesis and the widespread release of 02 by a little over 2 billion years ago is further confirmed by the inorganic rock record, in particular by the *banded iron formations* and *red beds.* The banded ironformations are composed of alternating layers of iron-rich and iron-poor sediments, most of which have ages between 2.0 and 2.4 billion years.

The iron-rich layers were laid down when iron that was dissolved in shallow ocean basins came in contact with 02. This converted the dissolved iron into rusted, oxygenrich iron, which then precipitated and settled to the ocean bottoms. There it accumulated with other sediments to become the iron-rich layers found today in the banded iron-formations. Some of these iron-rich layers are only a few millimeters thick, but often can be traced for hundreds of kilometers, indicating that iron was once dissolved over wide expanses in the ancient oceans.

The banded iron-formations are the source of many of the iron ore deposits that are mined today, as, for example, those of the Mesabi Range in Minnesota and the Marquette Range in Michigan.

The layering of the banded iron-formations suggests that periods when much oxygen-rich iron precipitated and accumulated on the floors of ocean basins alternated with periods when very little of it precipitated. This may have been due to periods of rapid growth of aerobic photosynthetic bacteria followed by periods of less rapid growth, with a consequent periodic rise and fall of the O_2 level in the atmosphere and oceans. It may also have been due to fluctuations in the rate at which iron was washed out of rocks by rain and became exposed to oxygen.

Although most banded iron-formations have ages between 2 and 2.4 billion years, some are considerably older as, for example, the Woman River and Isua formations. However, neither of these two formations can be unquestionably associated with early life. Hence, it is not clear whether the oxygen that precipitated the iron in those ancient rocks came from aerobic photosynthesizers or had a nonbiological origin, such as the splitting of water molecules by ultraviolet radiation from the sun.

If the O_2 was released by photosynthetic bacteria, life may have come into existence even earlier than 4 billion years ago. Very few of the banded iron-formations are younger than 2 billion years. Apparently by then so much O_2 had accumulated in the atmosphere that dissolved iron could no longer spread out extensively over the oceans.

The dissolved iron reacted with oxygen while it was still being carried by the rivers toward the oceans, and the resulting rusted iron was deposited (in the form of Fe_2O_3, called *hematite)* along with sand and gravel in relatively confined, onshore marine basins. Such deposits are known today as *red beds*.

The long history of banded iron-formations, followed by the relatively rapid transition to red bed deposits, indicates that the O_2 level in the Earth's ancient atmosphere rose exceedingly slowly for hundreds of millions of years. Only about 2 billion years ago did it reach levels at which it began to interact strongly and consistently with the environment.

In part, it rose so slowly because the iron that was dissolved in the oceans acted as a buffer that absorbed and removed the oxygen as it was released by photosynthesis. This, in turn, gave early life a sufficient period of grace to adapt to oxygen's high reactivity and to evolve aerobic respiration.

The banded iron-formations and red beds are not the only geological evidence of the slow rise of O_2 in our planet's atmosphere 2 billion years and more ago. For instance, the gold-uranium deposits of the Dominion Reef and Witwatersrand System of South Africa, which were laid down between 2.0 and 2.8 billion years ago, offer independent confirmation.

These deposits contain uranium and lead in the molecular forms UO_2 (called uraninite) and PbS (galena). In the presence of even small amounts of O_2 these molecules change rapidly into U308 (triuranium octoxide) and $PbSO_4$ (anglesite). The presence of UO_2 and PbS in the Dominion Reef and Witwatersrand rocks is a sure indication that the oxygen level up to about 2 billion years ago was indeed very low.

Eukaryotic Life

In response to the rise of O_2 in the Earth's atmosphere, aerobic respiration evolved and life made the transition from prokaryotic to eukaryotic forms. The oldest known evidence of this transition comes from a number of sites, such as the Bungle Bungle and Amelia dolomites of northern Australia and rocks in the southern Ural Mountains of Russia, with approximate ages of 1.5 billion years.

The fossil microbiotas preserved in these rocks are much more diverse than those found in older rocks. Those of the Bungle Bungle and Amelia deposits consist of associations between blue-green bacteria, growing in typical stromatolitic forms, and spherically shaped cells of various kinds.

The spherical cells are between 1 and 5 μm in diameter and occur either alone or in clusters of up to a hundred. Most important, some of them contain what appear to be small, membrane-enclosed structures that may be remnants of organelles, suggesting that these cells were eukaryotes. Furthermore, a group of four tightly bound fossilized cells, found in the Amelia dolomite, closely resembles the spores produced by the mitotic cell division of modern green algae.

Algae are eukaryotes and, hence, this ancient group of four cells is yet another indication that some of the organisms in these deposits may have been eukaryotes, though some researchers have questioned this interpretation. The microfossils found in the southern Ural Mountains have been identified by Russian and European paleontologists as originating from unicellular eukaryotic organisms that were buoyant and lived freely in the ancient sea as plankton.

As we go forward in time from 1.5 billion years ago, the fossil evidence for the existence of eukaryotic cells becomes progressively more common and convincing. The evidence comes from rock deposits of such widely dispersed areas as northern and southern Australia, southern India, arctic Canada, Montana, California, Utah, and Arizona. For instance, the California fossils are found in 1.2 to 1.4 billion-year-old strata of dolomite-the Beck Spring dolomite-of the Nopah Range, east of Death Valley.

As in the Bungle Bungle and Amelia dolomites, the fossilized microbiota of the Beck Spring dolomite consist of host stromatolites that have been formed by several different species of blue-green bacteria and harbor a variety of other kinds of cells. Some of these cells are almost certainly of eukaryotic origin. They are up to 60 p m in diameter, which is much larger than any of the earlier prokaryotic cells and comparable to the sizes of many modern eukaryotic cells.

Furthermore, they occur either singly or in clusters. Many contain minute dark spots and bands, suggestive of organelles. Some of them are smooth-surfaced, while others have surface ornamentations, such as granules and spines, similar to features found on contemporary green and yellow-brown algae.

In addition to having larger sizes and more complex internal and surface constructions than prokaryotes, many early eukaryotic organisms appear to have been plankton, as noted already in the case of the microfossils found in the southern Ural Mountains.

The evidence comes from the fact that the fossilized remains of many of these early eukaryotic microorganisms are found uniformly distributed in such sedimentary rocks as shale, sandstone, and limestone, suggesting that they became trapped as the original materials of these rocks precipitated toward the bottom of the sea.

That is what happens today to planktonic microorganisms. Some of these early eukaryotes may also have been motile, for the outer membranes of some of them contain small circular holes through which undulating flagella may have protruded. In contrast, most of the prokaryotes of that early time, such as those that formed the stromatolites, were members of benthic communities, meaning they were nonmotile and lived at the bottom of shallow seas.

Complex Multicelled Life

Life remained microscopic and mainly single-celled until about 670 million years ago. Then, within a relatively short span of time by geologic standards, macroscopic and multicelled marine forms appeared that bore a resemblance to modern jellyfish, worms, sea pens, soft corals, arthropods, and echinoderms. All of these marine creatures were soft-bodied; none possessed hard, mineralised shells or solid skeletons; and they inhabited shallow, near-shore marine environments. Some were free floating, others were anchored to the ocean floor, and still others crawled along in the mud.

It is interesting that with the arrival of complex multicelled life, stromatolites suffered a marked decline that has continued to the present. Apparently the blue-green bacteria that built the stromatolites could not stand up to competition with invertebrates. The age of the blue-green bacteria came to a close.

The first convincing evidence of the existence of Precambrian animal life came in 1930 when the German paleontologist Ernst J. G. GOrich found traces of multicellular organisms in roughly 600 million year old rocks in Namibia, Africa. Then, in the mid-1940s, the Australian geologist Reginald C. Sprigg discovered additional and more extensive evidence in the Ediacara Hills some 400 km north of Adelaide, Australia.

There, a band of rock outcrops, exposed over a 140-km-long stretch and approximately 670 to 570 million years old, contains a variety of exceptionally well preserved fossils of multicelled, soft-bodied marine organisms. Martin F. Glaessner, of the University of Adelaide, has identified nearly two dozen species among the Ediacara fossils, as well as numerous surface tracks and shallow burrows made by the animals.

Since the 1940s, other fossilized marine organisms of similar constitution and age have been found in rock deposits over much of the globe: in China, Siberia, the western U.S.S.R., Scandinavia, England, and Canada. They are collectively referred to as the *Ediacara fauna,* and the period of their existence-from about 670 to 570 million years ago - has become known as the *Ediacarian period.*

The discovery of the Ediacara fauna calls for a redefinition of the traditional geologic reckoning of time. The Phanerozoic eon should be extended back in time to 670 million years ago, to include the Ediacarian period. The Ediacarian period becomes then the oldest of twelve geologic periods of the Phanerozoic eon, predating the Cambrian.

In keeping with this change, the Precambrian eon needs to be renamed the Prephanerozoic eon and to end 670 million years ago, when visible life (complex macroscopic and multicellular life, visible with the unaided eye) began to appear. Preston Cloud has proposed such changes in the geologic time scale. It remains to be seen whether they will be adopted by the geologic community.

FROM FERMENTATION TO RESPIRATION

The previous section followed the evolution of early life as it is revealed by fossils and mineral deposits in ancient rocks. Additional insights into this evolution can be gained by studying the metabolic pathways and other biochemical processes in contemporary cells. Let us start by exploring the ways cells obtain energy.

During normal activity the cells of our bodies obtain energy by aerobic respiration. We inhale atmospheric oxygen; the blood stream carries it to our cells; and in the cells carbohydrates, in the form of glucose ($C_6H_{12}O_6$), are broken down and combined with the oxygen to form carbon dioxide and water, with the simultaneous release of energy.

In contrast, during brief spurts of strenuous exercise that leave us breathless our cells gain energy quite differently. There is not sufficient time to bring enough O_2 to the cells, especially to the muscle cells, for carrying the respiratory reactions to completion. Our cells revert then to a primitive metabolism that does not use oxygen: glucose is converted to lactic acid.

This *anaerobic* metabolism releases only a relatively small amount of energy compared to respiration and it corresponds in all important respects to bacterial fermentation. The only significant difference between them is that in our cells the end product is aways lactic acid, while in bacteria it may also be alcohol, vinegar, or some other simple organic compound.

It seems our cells "remember" their bacterial ancestry. Normally they meet their energy needs by aerobic respiration, but when there is a lack of oxygen, they fall back upon fermentation, the metabolism of the earliest prokaryotes. However, our cells can function in this mode for only a minute or two at a time. Fermentation does not free enough energy to sustain the needs of large, multicelled bodies like ours.

Fermentation

Let us take a closer look at fermentation. It consists of two separate reaction sequences that, like all biochemical reactions, are catalyzed by enzymes. In the first reaction sequence, called *glycolysis,** each glucose molecule is broken down into two pyruvate molecules and energy is released:

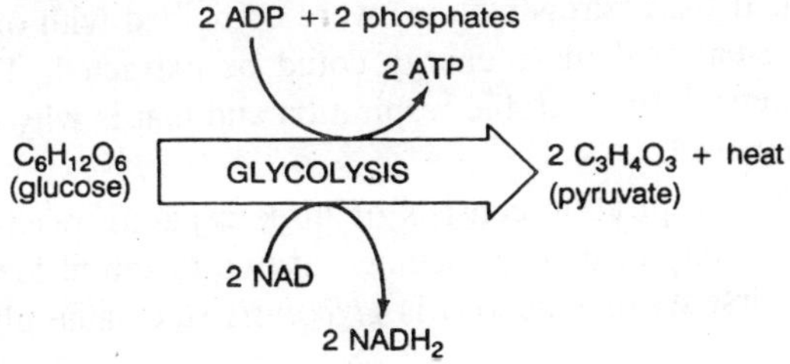

2 ADP + 2 phosphates
2 ATP
$C_6H_{12}O_6$ (glucose)
GLYCOLYSIS
2 $C_3H_4O_3$ + heat (pyruvate)
2 NAD
2 $NADH_2$

Some of the energy becomes stored in two molecules of ATP, and the rest is lost as heat. The four hydrogen atoms (H), which the two pyruvates are short in comparison to the original glucose, are picked up by two carrier molecules called NAD *(nicotinamide adenine dinucleotide)*. Each of the NADs picks up two hydrogen atoms and is thereby converted to $NADH_2$.

The $NADH_2$ molecules keep their hydrogen atoms only temporarily. They enter the second reaction sequence of fermentation (which has no particular name) and react with the pyruvates. The $NADH_2$ molecules are stripped of their Hs and converted back to NADs, ready to participate in further reactions of glycolysis.

The hydrogen atoms and the pyruvates combine to yield any number of different end products, depending on the species of bacteria involved: ethanol (drinking alcohol, C_2H_6O), acetic acid (vinegar, $C_2H_4O_2$), lactic acid ($C_3H_6O_3$), butyric acid ($C_4H_8O_2$), or some other organic compound of similar complexity. For instance, the reactions that yield drinking alcohol and lactic acid may be summarised as follows:

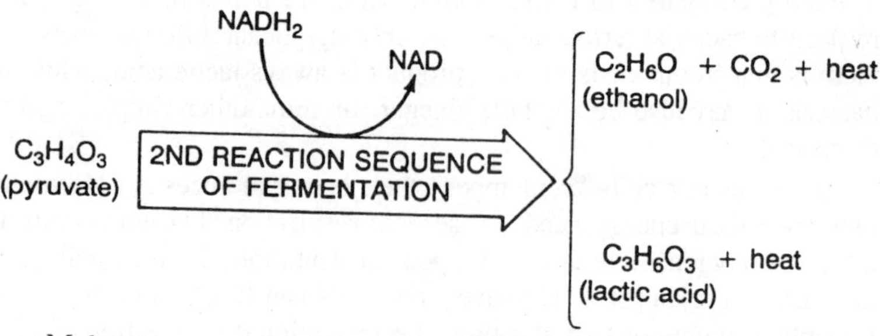

Molecular oxygen, O_2, does not participate in any of these reactions. That is why the French chemist Louis Pasteur referred to fermentation as "the consequence of life without air."

Aerobic Respiration

Fermentation is a very wasteful process because its end products -lactic acid, ethanol, and so forth—still contain most of the energy originally present in glucose. All of these end products are rich in hydrogen. If the hydrogen atoms were combined with oxygen to form water, a great deal more energy could be extracted. That is exactly what happens during aerobic respiration and that is why this metabolic pathway evolved.

Aerobic respiration consists of three separate reaction sequences that are carried out in sequence and at two different locations- in the cell. The first set of reactions is *glycolysis,* the same glycolysis as in bacterial fermentation. It takes place in the cytoplasm of the cell. The second and third reaction sequences are the *citric acid cycle* and the *respiratory chain.*

Bacteria capable of aerobic respiration carry these reactions out on the inner surfaces of their enclosing membranes. Eukaryotic cells carry them out on the inner membranes (the cristae) of the mitochondria, the organelles described as the cell's "powerhouse" in the Introduction to Part Two.

During glycolysis each glucose molecule is broken down into two pyruvates and, simultaneously, two molecules of ATP and two of NADH2 are produced. The ATPs store useful energy and are applied much as they are in fermenting cells. However, the pyruvates and $NADH_2$ follow rather different reaction paths.

The two pyruvates enter one of the mitochondria of the cell and start passing through the citric acid cycle. With the participation of 3 H_2O molecules, each pyruvate is broken down into 3 CO_2 molecules and 10 Hs. The Hs are carried away by 5 carrier molecules, 4 of which are NADs and one is a similar kind of molecule called *flavin adenine dinucleotide* or FAD. In addition, one ATP molecule is formed:

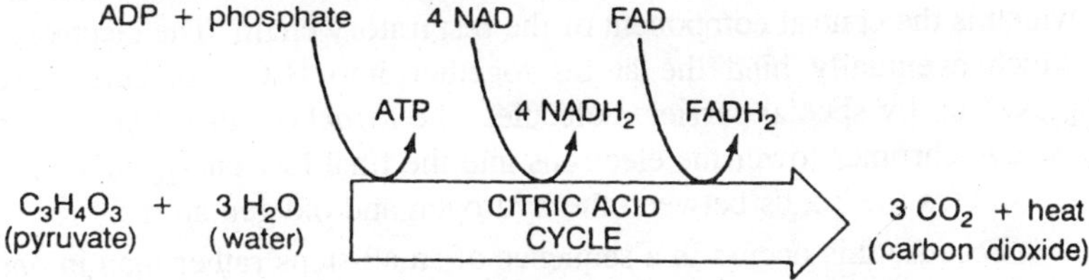

No oxygen is used during the citric acid cycle. Much of the energy that was present in the pyruvate is now stored in the $NADH_2$s and the rest has been converted to heat.

The 2 $NADH_2$s formed during glycolysis, the 8 formed during the citric acid cycle (4 from each of the 2 pyruvates), and the 2 $FADH_2$s, also formed during the citric acid cycle, now enter the respiratory chain. Only now does oxygen enter into the reactions. The reactions carried out by the respiratory chain are known as *oxidative phosphorylation* iecause they generate ATP by adding a phosphate to ADP and use oxygen.

Each NADH2 and $FADH_2$ gives up its hydrogen atoms and becomes once again an empty carrier molecule, NAD and FAD, ready for renewed use. The hydrogen atoms combine with oxygen to form water and a great deal of energy is released, some of which is stored in molecules of ATP:

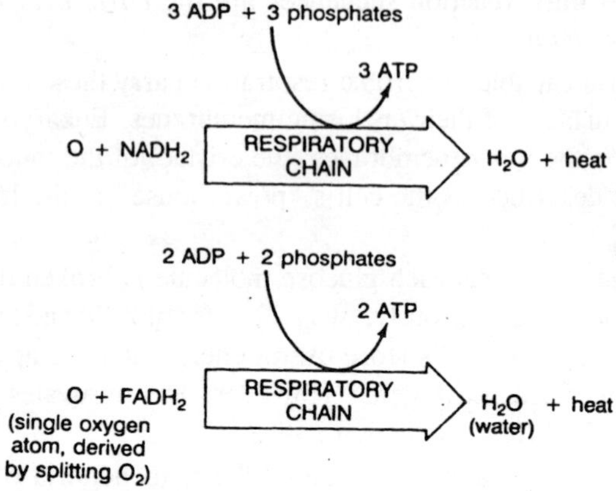

The chemical reactions between hydrogen and oxygen atoms to form water molecules release enough energy to make them potentially explosive. To avoid this danger, the reactions of the respiratory chain must be carried out slowly and in a controlled fashion.

This is done by the electron-transport chain, mentioned earlier, which is the central component of the respiratory chain. The electrons, which eventually bind the atoms together into H_2O molecules, are picked up by special carrier molecules, the *cytochromes*. Step-by-step the cytochromes lower the electrons into the final low-energy state that constitutes the bonds between the hydrogen and oxygen atoms.

Because this occurs in a sequence of small steps rather than in one big jump, the energy comes off gradually and is usable for forming ATPs. No harm is done to the cell.

The need for releasing the energy in a controlled, step-by-step manner becomes clear when we look at an example without such control, namely the burning of wood. Wood consists largely of glucose molecules bonded together into cellulose.

When wood burns, the glucose molecules combine with O_2 from the air to produce CO_2 and H_2O. However, unlike aerobic respiration, no electron-transport chain is involved and the energy is released as rapidly as 02 can get to the glucose.

We feel that energy as the heat coming off a wood fire. If the same uncontrolled process occurred in our cells, they would be immediately destroyed.

Now let us reconsider the reactions of aerobic respiration,

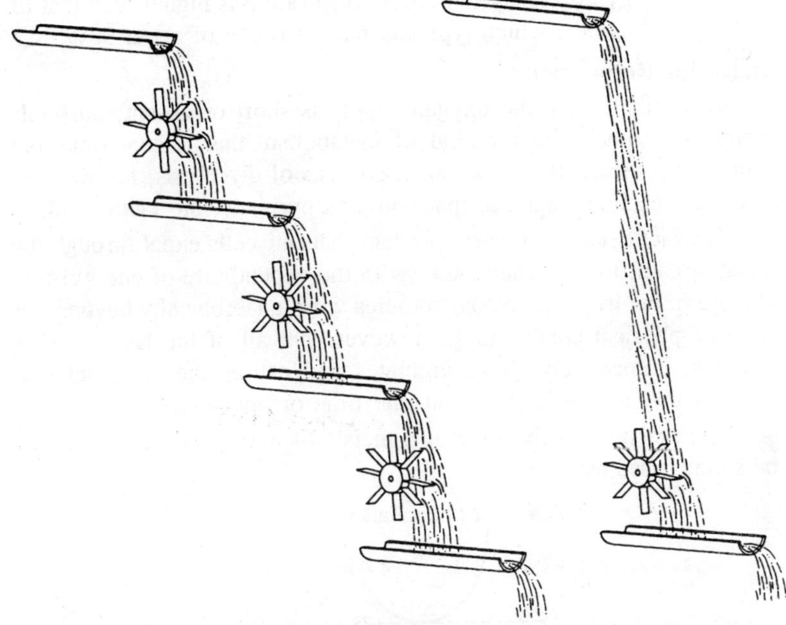

Figure 5.3 : Using the energy of running water. If the water falls over a series of water wheels, as shown on the left, each wheel receives only a fraction of the total energy. Consequently, the wheels turn slowly and gently. If the water falls over the entire height at once, as shown on the right, the single wheel receives the full impact and turns rapidly and forcefully.

disregarding all details:

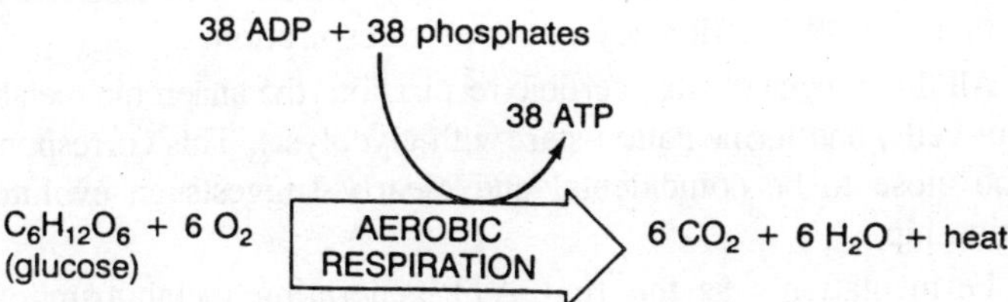

It turns out that roughly 40% of the total energy liberated is stored in ATP molecules and the rest is given up as heat. The heat energy is of no further use to the cell, except for helping to maintain body temperature, and is released to the environment.

The energy stored in molecules of ATP, on the other hand, is useful energy. The ATPs deliver it to wherever it is needed by the cell, such as to the ribosomes for assembly of peptides, to the nucleus for replication of DNA, to the cell membrane for active transport of molecules in and out of the cell, and to hundreds of other places of cell activity. Incidentally, the 40% energy efficiency of aerobic respiration is higher than that of combustion engines, which typically have a rating of about 30%.

Anaerobic Respiration

Recall that when the oxygen supply is short or absent, our cells switch to a much simpler kind of metabolism that is anaerobic and similar to bacterial fermentation. It consists of glycolysis, followed by a second reaction sequence that converts pyruvate into lactic acid.

The lactic acid is a waste product that our cells expel through the blood stream, to be turned back (with the expenditure of energy) into glucose in the liver. When our muscles work anaerobically beyond our level of physical conditioning, however, not all of the lactic acid is removed immediately. It accumulates and causes the sore and stiff muscles that we all have felt at one time or another.

The reactions of the anaerobic metabolism of eukaryotic cells may be summarised as follows:

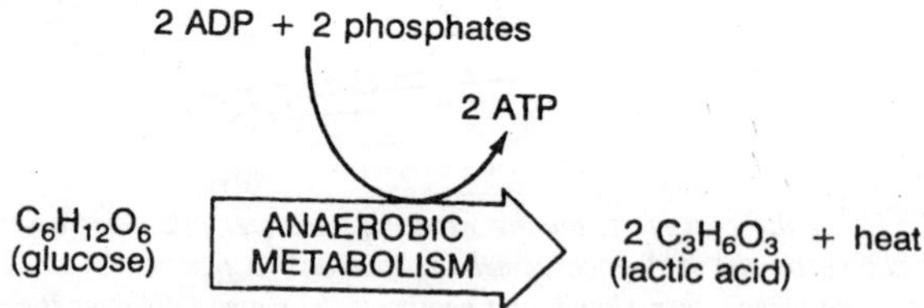

As in bacterial fermentation, this is a very inefficient way of harnessing energy. Only two molecules of ATP are produced per glucose molecule, which corresponds to an energy efficiency of a mere 2%, far less than the 40% efficiency of aerobic respiration.

All three metabolisms -aerobic respiration, the anaerobic metabolism of our cells, and fermentation-start with glycolysis. This correspondence is too close to be coincidental and clearly suggests an evolutionary relationship.

Fermentation was the first ATP-generating metabolism of life. Gradually, other pathways were added as the early prokaryotes kept adapting to the everchanging environment and competed in the search for new sources of raw materials and energy.

Eventually, as the O_2 level in the atmosphere of our planet began to rise, the citric acid cycle and the respiratory chain evolved and aerobic respiration came into existence.

It allowed life to exploit the very substance that had been poison to it earlier, and to extract enormously more energy from glucose than by fermentation alone.

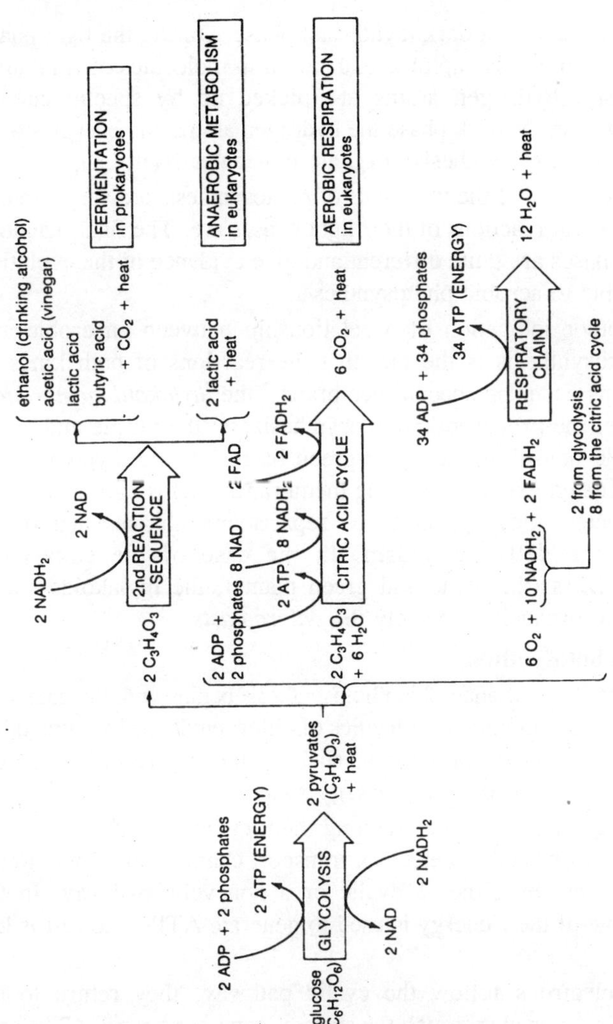

Figure 5.5: Comparison of the three metabolic pathways by which cells convert energy stored in glucose into energy of ATP All three pathways—the fermentative pathways of prokaryotes and the anaerobic and aerobic pathways of eukaryotes—start with glycolysis, which breaks down glucose into pyruvate and produces two ATP per glucose molecule.

THE EVOLUTION OF PHOTOSYNTHESIS

As in the case of fermentation and aerobic respiration, the major stages in the evolution of photosynthesis are still apparent in the ways in which contemporary organisms harness the energy of sunlight. The purple and green sulfur bacteria and the purple nonsulfur bacteria harness it *anaerobically;* the bluegreen bacteria, algae, and green plants harness it *aerobically*. Both kinds of photosynthesis take place in two phases, a light (photo) phase and a dark (synthesis) phase. During the light phase the energy of sunlight is captured and stored in molecules of ATP and, simultaneously, hydrogen atoms are picked up by special carrier molecules. During the dark phase the hydrogen atoms and energy stored in ATPs are used to synthesize organic molecules from CO_2.

The dark phases of the two kinds of photosynthesis are very similar, but only the initial reactions of their light phases are. The final reactions of the light phases are quite different and give evidence of the evolution from anaerobic to aerobic photosynthesis.

Yet another indication of a relationship between anaerobic and aerobic photosynthesis is the fact that the reactions of both kinds of photosynthesis occur on special membranes, the *thylakoid membranes*. In the case of the prokaryotic photosynthesizers (the purple and green sulfur, purple nonsulfur, and blue-green bacteria), the thylakoids are either invaginated extensions of the membranes enclosing the cells or, in some species, they appear to be separate membranes that reside directly in the cells cytoplasm. In the case of the eukaryotic photosynthesizers (the algae and green plants), the thylakoids reside within special organelles, namely the *chloroplasts*.

Anaerobic Photosynthesis

The light phase of anaerobic photosynthesis begins with the gathering of the energy of sunlight by molecules of chlorophyll (or by some other molecules with similar properties, such as the carotenoids) and the boosting of electrons to a high energy state.

Special carrier molecules (among them cytochromes) pick up the energetic electrons and via electron-transport chains return them to the low energy state by either a cyclic or a noncyclic pathway. In the process, some of their energy is used to generate ATPs, the rest is lost as heat.

If the electrons follow the cyclic pathway, they return to the chlorophyll and recycle through the system, generating more ATPs each time they complete a cycle. This pathway is called *cyclic photophosphorylation* because it is cyclic, uses the energy of photons, and generates ATPs by adding phosphate molecules to ADPs.

If the electrons follow the noncyclic pathway, they combine with positive hydrogen ions to form neutral hydrogen atoms, which are picked up by special carrier molecules called NADP *(nicotinamide adenine dinucleotide phosphate)*. Each NADP picks up two Hs and is thereby converted to $NADPH_2$. The molecular structure of NADP is identical to that of the NAD of aerobic respiration except for an extra phosphate,

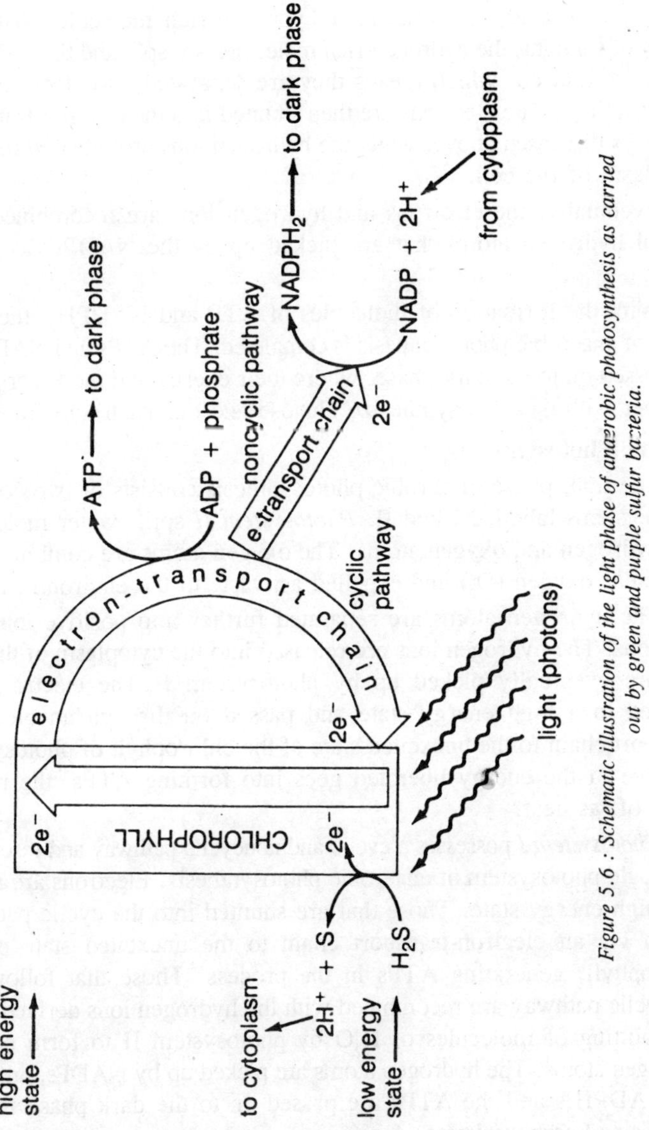

Figure 5.6 : Schematic illustration of the light phase of anaerobic photosynthesis as carried out by green and purple sulfur bacteria.

indicating that the two kinds of molecules have a common ancestry.

The electrons and hydrogen ions used in anaerobic photosynthesis may come from various sources. The green and purple sulfur bacteria derive them from the hydrogen atoms of H_2S. The purple nonsulfur bacteria obtain them from the hydrogen atoms of H_2, ethanol, lactic acid, pyruvic acid, or some other hydrogen-rich molecule. With the energy of sunlight, the hydrogen-rich molecules are split and the hydrogen atoms are ionised, which means they are separated into electrons and hydrogen ions. The electrons are then shunted into the electron-transport pathways discussed above, while the hydrogen ions are released into the cytoplasm of the cell.

Eventually, the electrons and hydrogen ions are recombined into neutral hydrogen atoms that are picked up by the NADPs, as noted above.

With the formation of molecules of ATP and $NADPH_2$, the light phase of anaerobic photosynthesis is completed. The ATPs and $NADPH_2$s are passed on to the dark phase, where their energy and the Hs are used together with CO_2s to synthesize glucose and other carbohydrates.

Aerobic Photosynthesis

The light phase of aerobic photosynthesis consists of two coupled photosystems labelled I and II. *Photosystem II* splits water molecules into hydrogen and oxygen atoms. The oxygen atoms are combined into molecular oxygen (O_2) and expelled as waste to the environment.

The hydrogen atoms are separated further into positive ions and electrons. The hydrogen ions are released into the cytoplasm of the cell and are eventually picked up by photosystem I. The electrons are boosted to a high energy state and passed on through an electron-transport chain to the unexcited state of the chlorophyll of photosystem I. Some of the energy liberated goes into forming ATPs, the rest is given off as heat.

Photosystem I possesses a cyclic and noncyclic pathway and resembles the single photosystem of anaerobic photosynthesis. Electrons are raised to a high energy state. Those that are shunted into the cyclic pathway return via an electron-transport chain to the unexcited state of the chlorophyll, generating ATPs in the process. Those that follow the noncyclic pathway are recombined with the hydrogen ions derived from the splitting of molecules of H_2O by photosystem II to form neutral hydrogen atoms. The hydrogen atoms are picked up by NADPs, and both the $NADPH_2$s and the ATPs are passed on to the dark phase for the synthesis of carbohydrates.

As in fermentation and aerobic respiration, the similarities between anaerobic and aerobic photosynthesis are too close to be accidental and thus manifest an evolutionary relationship. Anaerobic photosynthesis, with its single photosystem, developed first. In the course of time, a second photosystem emerged with a slightly modified chlorophyll that

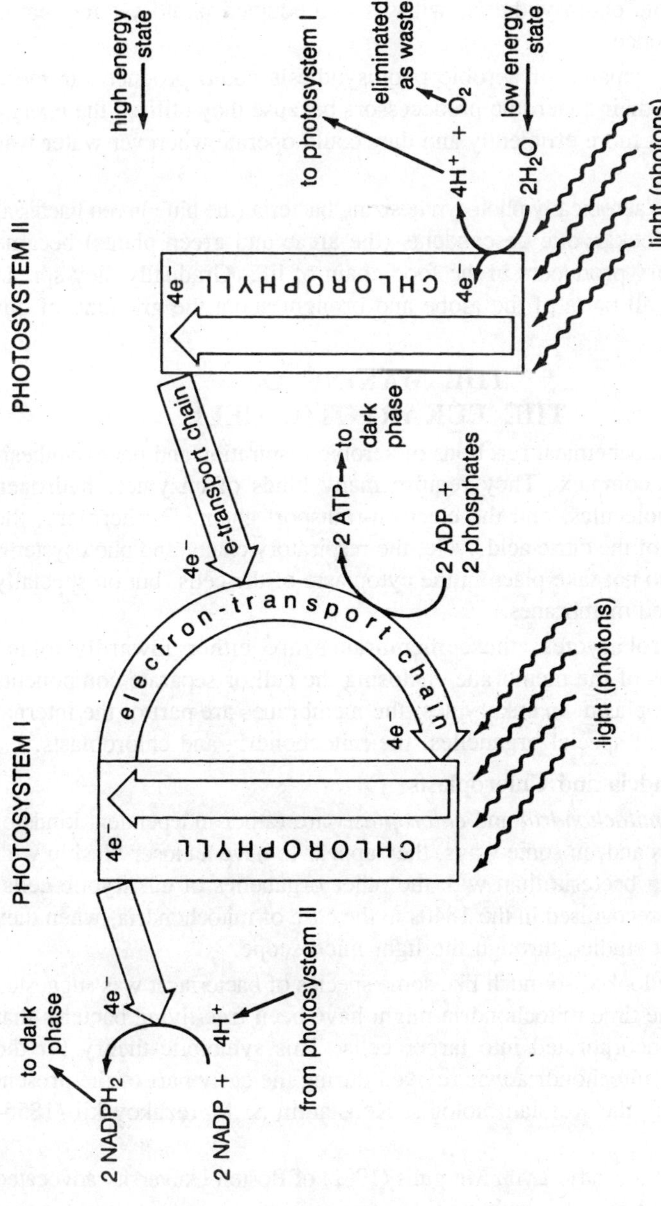

Figure 5.7 : Schematic illustration of the light phase of aerobic photosynthesis. (This diagram is to be read from right to left, starting with photosystem II.)

absorbed light of shorter wavelengths and could split water molecules. A virtually inexhaustible source of hydrogen atoms was thereby tapped and aerobic photosynthesis, with its two coupled photosystems, came into existence.

Cells capable of aerobic photosynthesis could produce far more food than their anaerobic predecessors because they utilised the energy of sunlight more efficiently and they could operate wherever water was available.

Thus, aerobically photosynthesizing bacteria (the blue-green bacteria) and their eukaryotic descendents (the algae and green plants) became the primary producers in the food chain of life. Gradually they spread to nearly all parts of the globe and brought about the greening of our planet.

THE MAKING OF THE EUKARYOTIC CELL

The biochemical reactions of aerobic respiration and photosynthesis are quite complex. They require many kinds of enzymes, hydrogen carrier molecules, and the electron-transport chain. Furthermore, the reactions of the citric acid cycle, the respiratory chain, and photosystems I and II do not take place in the cytoplasm of the cells, but on specially constructed membranes.

In prokaryotes, these membranes are either inwardly-folded extensions of the membrane enclosing the cell or separate components in the cytoplasm. In eukaryotes, the membranes are part of the internal structure of special organelles, the mitochondria and chloroplasts.

Mitochondria and Chloroplasts

The *mitochondria* and *chloroplasts* are rather independent kinds of organelles and, in some ways, they appear to have a closer kinship with free-living bacteria than with the other organelles of eukaryotic cells. This was recognised in the 1840s in the case of mitochondria, when they were first studied through the light microscope.

They looked so much like some species of bacteria, it was suggested that at one time mitochondria might have been free-living bacteria that became incorporated into larger cells. This symbiotic theory for the origin of mitochondria was revived during the early part of the present century by the Russian biologist Konstantin S. Merezhkovskii (1855-1921).

More recently, Lynn Margulis (1982) of Boston University advocated the theory again, for both mitochondria and chloroplasts. She noted that

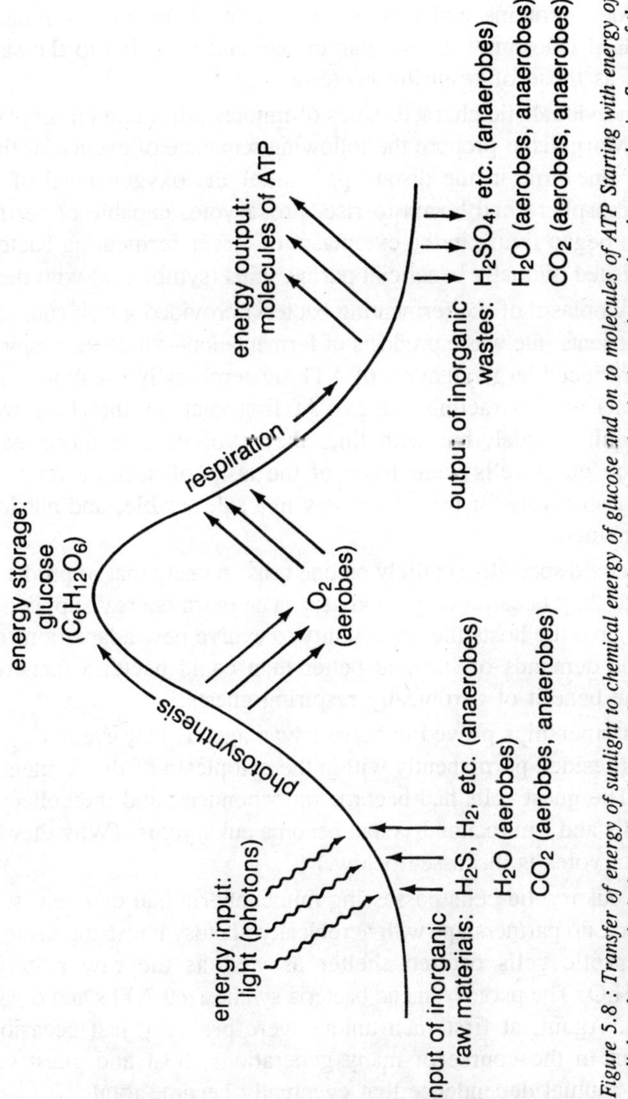

Figure 5.8 : Transfer of energy of sunlight to chemical energy of glucose and on to molecules of ATP Starting with energy of sunlight, photosynthetic bacteria, algae, and plants assemble inorganic molecules into glucose by photosynthesis. Some of the energy of sunlight is thereby stored in the chemical bonds of glucose. Glucose that enters the food chain (for example, as food for animals) is broken down by respiration.

both organelles possess their own DNA, mRNA, tRNA, and ribosomes, and that they manufacture some of their own proteins, much as bacteria do. Their reproduction goes on quite independently of the reproduction of the eukaryotic cells within which they reside, and it happens by a division process that is similar to the binary cell division of bacteria. (However, mitochondria and chloroplasts are absolutely dependent upon nuclear-coded proteins and cannot survive by themselves.) Finally, mitochondrial ribosomes are similar in size and sensitive to the same antibiotics as those of respiring bacteria.

The individualistic characteristics of mitochondria and chloroplasts prompted Margulis to propose the following sequence of events for their origin. At one time in the distant past, after the oxygen level of the Earth's atmosphere had begun to rise, prokaryotes capable of aerobic respiration began living in the cytoplasm of larger fermenting bacteria and established mutually beneficial partnerships (symbioses) with them.

The cytoplasm of the fermenting bacteria provided a rich source of organic nutrients -the waste products of fermentation—while the respiring bacteria produced large amounts of ATP by aerobically breaking down the nutrients and extracting energy. At first such partnerships were probably quite casual, but with time they evolved into more stable unions. The guest cells were freed of the tasks of seeking food and fending for survival, for they lived now in a safe, stable, and nutrient-rich environment.

They could specialize entirely on one task - namely that of producing ATPs - and they became very good at it. The extra energy supplied by the ATPs gave the hosts the opportunity to evolve new adaptations and to meet the demands of survival better than could bacteria that lived without the benefit of aerobically respiring guests.

The partnerships proved to be so advantageous that eventually the guest cells resided permanently within the cytoplasm of the fermenting bacteria. The guest cells had become mitochondria, and the collection of host cells and , mitochondria had become eukaryotes. (Why they are called eukaryotes is discussed below.)

Once eukaryotic cells possessing mitochondria had evolved, some of them took up partnerships with aerobically photosynthesizing bacteria. The eukaryotic cells offered shelter as well as the raw materials $CO2_2$and H_2O. The photosynthetic bacteria synthesized ATPs and organic molecules. Again, at first such unions were probably just occasional events. But in the course of many generations, host and guest cells evolved a mutual dependence that eventually became total.

The photosynthetic bacteria had become chloroplasts and taken up permanent residence within their hosts. The modern, photosynthetic eukaryotic cell had evolved. (All eukaryotic photosynthetic cells - the algae and the cells of green plants -possess mitochondria in addition to chloroplasts, and they are able to carry out aerobic respiration.)

The symbiotic theory for the origin of mitochondria and chloroplasts

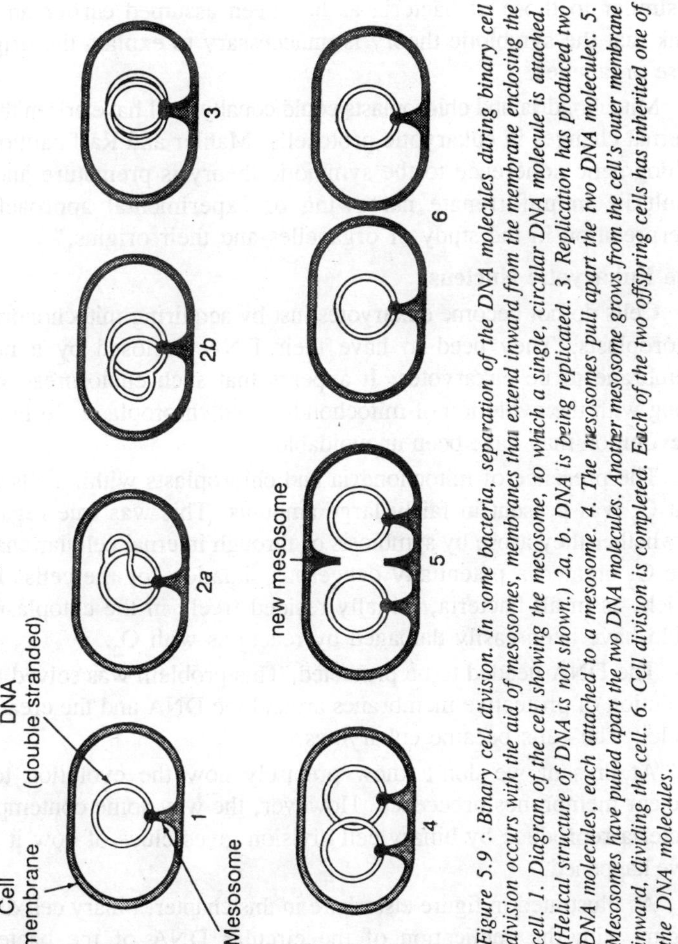

Figure 5.9 : Binary cell division. In some bacteria, separation of the DNA molecules during binary cell division occurs with the aid of mesosomes, membranes that extend inward from the membrane enclosing the cell. 1. Diagram of the cell showing the mesosome, to which a single, circular DNA molecule is attached. (Helical structure of DNA is not shown.) 2a, b. DNA is being replicated. 3. Replication has produced two DNA molecules, each attached to a mesosome. 4. The mesosomes pull apart the two DNA molecules. 5. Mesosomes have pulled apart the two DNA molecules. Another mesosome grows from the cell's outer membrane inward, dividing the cell. 6. Cell division is completed. Each of the two offspring cells has inherited one of the DNA molecules.

is supported further by the fact that partnerships between different kinds of organisms are found commonly throughout the contemporary biological world, Quite often such partnerships lead to complete dependence between the participating organisms. There is no reason to think that such partnerships did not also exist in the past and that they significantly affected the evolution of life.

However, a word of caution is in order. Although the symbiotic theory for the origin of mitochondria and chloroplasts is supported by much evidence and most biologists accept it as plausible, some do not. For example, Henry R. Mahler and Rudolf A. Raff of Indiana University assert that the genetic systems of mitochondria and chloroplasts are not as similar to those of bacteria as had been assumed earlier and they think that the symbiotic theory is unnecessary to explain the origin of these organelles.

Mitochondria and chloroplasts could equally well have arisen through internal changes in eukaryotic protocells. Mahler and Raff caution that a "dogmatic adherence to the symbiotic theory is premature and may result in an unfortunate narrowing of experimental approach and interpretation in the study of organelles and their origins."

The Eukaryotic Nucleus

Cells do not become eukaryotes just by acquiring mitochondria and chloroplasts. They need to have their DNA enclosed by a nuclear membrane to be eukaryotes. It appears that such enclosures evolved along with the evolution of mitochondria and chloroplasts. In fact, this coevolution may have been unavoidable.

The presence of mitochondria and chloroplasts within cells meant that O_2 was present in fairly large amounts. This was true regardless of whether they arose by symbiosis or through internal cellular changes. The O_2 created a potentially dangerous situation for the cells' DNA, which, as in all bacteria, initially resided freely in the cytoplasm and could have been easily damaged by reactions with O_2.

The DNA needed to be protected. This problem was solved by the evolution of protective membranes around the DNA and the creation of nuclei. The cells became eukaryotes.

At present we don't know precisely how the evolution toward nuclear membranes proceeded. However, the way some contemporary bacteria reproduce by binary cell division gives clues of how it might have happened.

As illustrated in figure elsewhere in this chapter, binary cell division begins with the replication of the circular DNA of the bacterium, temporarily giving the cell two identical pieces of DNA. Simultaneously, in some bacteria portions of the cell's enclosing membranes grow inward and attach themselves to the two pieces of DNA.

The inward-growing membranes are called *mesosomes,* from *meso-* meaning "middle" and *-some* meaning "body." Gradually the cell membrane between the two mesosomes grows larger and the mesosomes separate in opposite directions, pulling the two pieces of DNA apart.

At the same time, an additional mesosome grows inward somewhere between the two pieces of DNA and divides the cell into two offspring cells, with each receiving one piece of the DNA. The cell division is thereby completed. It is easy to imagine how in some ancestral bacteria the need to protect the DNA from O_2 might have led to the evolution of larger mesosomes that eventually enveloped the DNA and thus became the nuclear membranes of eukaryotic cells.

EUKARYOTIC CELL DIVISIONS: MITOSIS AND MEIOSIS

However the nucleus of eukaryotes evolved and for whatever reason, one thing is certain: it changed the course of biological evolution. It changed it for two reasons. First, by enclosing the DNA in a membrane-bound structure and separating it from the chemical processes in the rest of the cell, the nucleus provided an environment in which the DNA could evolve to much greater lengths than was possible before.

This meant that eukaryotic cells could carry considerably more genetic information and had the potential for evolving much more complex adaptations than did prokaryotes. Second, the nucleus permitted a new kind of organisation and mechanical handling of the DNA, which led to the development of two new kinds of cell divisions-mitosis and *meiosis*. Mitosis is involved in the nonsexual reproduction of eukaryotic cells, while meiosis is the basis of their sexual reproduction.

Eukaryotic DNA

Eukaryotic DNA is typically ten to many hundreds of times longer than bacterial DNA. It varies from about 1 cm in very simple eukaryotes, such as yeasts and sponges, to about a meter or more in the case of humans, many other animals, and many plants.

Eukaryotic DNA is closely associated with a number of different proteins. Some of the proteins act as enzymes and help in the repair, replication, and transcription of the DNA. Others act as structural proteins and help keep the long strands of eukaryotic DNA organised into manageable units.

The latter kinds of proteins are called *histones*. Histones are rich in positively charged amino acids and are arranged into roundish bodies known as *nucleosomes*. The positive charges of the histones cancel the negative charges carried by DNA on the phosphates of its backbone and greatly increase the degree to which DNA can be compacted in the cell nucleus.

Each strand of DNA is associated with a sequence of tens to hundreds of thousands of nucleosomes, around which it is wrapped much like a long piece of thread might be wrapped around a sequence of spools.

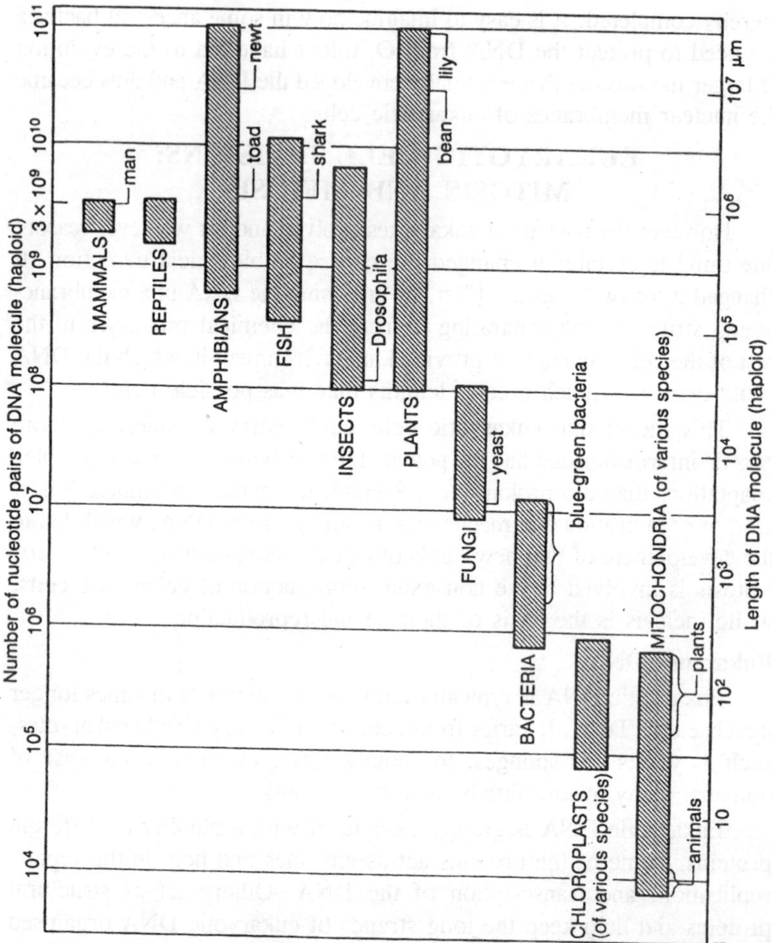

Figure 5.10 : Length of DNA molecules. Note that the DNA of some plants and amphibians is longer than that of bacteria by a factor of 10^4 to 10^5. The total length of human DNA is approximately 10^6 μm or 1 m, which is equivalent to roughly 3×10^9 nucleotide pairs or 10^9 codons. Obviously, the length of DNA carried by an organism is not necessarily related to its phenotypic complexity We may conclude that DNA does not carry useful information over its entire length. Much of its coding is "nonsense."

Another feature unique to eukaryotic DNA is that it never exists in one piece, as does bacterial DNA, but always in two or more pieces. For example, there are 8 pieces of DNA in the cells of the fruit fly *Drosophila*, 48 in those of chimpanzees, gorillas, and orangutans, and 46 in those of humans. Apparently, only by dividing their DNA into several separate pieces are eukaryotic cells able to handle the DNAs great length. During the initial stages of mitosis and meiosis (just after replication of DNA is completed, see below) the various pieces of DNA, together with their histones and other proteins, undergo several successive levels of folding and coiling.

Finally they become the short, stubby bodies that are familiar from photographs made through the light microscope. In 1888, the German physiologist Wilhelm von Waldeyer-Hartz named these bodies *chromosomes,* meaning "coloured bodies," for their affinity for certain dyes.

Mitosis

Let us now turn to *mitosis,* the form of cell division by which eukaryotic cells reproduce without sex. Examples of mitosis are the repeated division of a fertilized egg and its development into an adult, the replacement of old or damaged cells in our bodies, the growth of plants, and the multiplication of yeast cells.

Strictly speaking, mitosis refers only to the processes involved in the division of the cell nucleus: the replication of DNA and the equal distribution of the two sets of DNA among the offspring cells. The processes that affect the cytoplasm of the cell are called *cytokinesis,* which literally means "cell motion."

Cytokinesis includes cleavage of the cell and the distribution of cytoplasmic constituents and organelles among the two offspring cells.

Mitosis and cytokinesis are dynamic and complicated processes, whose beauty is difficult to portray in words or static pictures. Both processes occur in innumerable variations and are never exactly alike from one cell division to the next.

In particular, they are somewhat different in plant and animal cells, mainly because plant cells are surrounded by rigid cellulose walls and animal cells are not. Still, the processes of mitosis and cytokinesis of all eukaryotic cells share certain essential features that permit their description in a schematic and generally valid manner.

Traditionally, the processes of mitosis and cytokinesis have been divided into four steps or phases, although in reality they are one continuous sequence of events. The four phases are *pro-, meta-, ana-,* and *telophase.* In addition there is a fifth phase *-interpbase-which* refers to the time interval between successive cell divisions. The phases are described in figure elsewhere in this chapter and below.

1. *Interphase* is the period between successive cell divisions. Except in very rapidly dividing cells, it is the longest of the five phases and

The two strands of the DNA molecule wind around each other in a spiral or helical fashion.

During *interphase*, DNA is *replicated*. Two "sister" DNA molecules are created.

Each DNA double helix wraps around histone complexes like a thread around a sequence of spools.

During *prophase*, each DNA and its histones fold repeatedly, creating short, stubby *chromosome bodies*.

Each chromosome consists of two identical halves, the *chromatids*, held together at the *centromere*. (The two "sister" chromatids correspond to the two "sister" DNA molecules created by replication.)

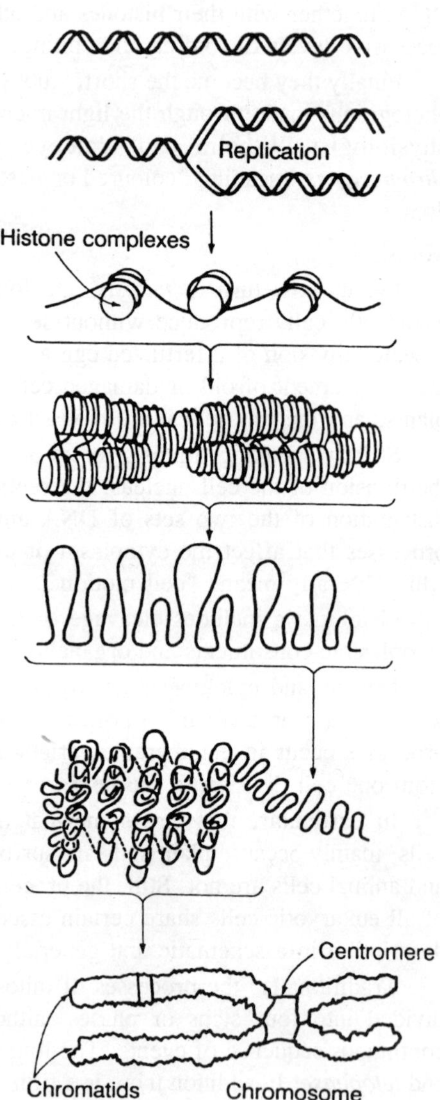

Figure 5.11 : Schematic illustration of the different levels of packing of DNA into highly condensed chromosome bodies during eukaryotic cell division.

is sometimes called the resting stage of the cell, even though during this phase the cell does everything but rest.

It actively carries out its various metabolic processes needed for growth and self-maintenance as well as its specialised tasks, such as contraction if it is a muscle cell, production of hormones if it is an endocrine cell, and ingestion and destruction of bacteria if it is a white blood cell. Interphase is also the period during which DNA is replicated.

2. During *prophase* the strands of DNA, which up to now have remained dispersed in the nucleus, coil into the short, stubby chromosome bodies described above. Because each DNA has been replicated during interphase, each of the chromosome bodies consists of two identical halves, called *chromatids.*

The two chromatids of each chromosome are held together at a constricted region known as the *centromere.* Also during prophase the nuclear membrane begins to break up into fragments and the *spindle apparatus* forms.

The spindle apparatus consists of hundreds of thin fibers, some of which radiate outward from two clearly distinguishable *polar regions,* while others radiate away from the centromeres of the chromosomes. Each polar region is centered on a pair of tiny structures, the *centriole pairs.* Changes in the lengths of the spindle fibers are responsible for the dynamic processes of the next two phases.

3. With the onset of *metaphase,* the disintegration of the nuclear membrane is completed and the membrane fragments disperse into the cytoplasm. There is a flurry of activity as the chromosomes, guided by the spindle fibers, settle down in the middle of the cell with their centromeres aligned along a plane perpendicular to the polar axis.

4. During *anaphase,* the cell lengthens in the direction of the polar axis and the chromatids, which have been held together at their centromeres, break apart. One set of chromatids—which constitutes one complete set of chromosomes—moves toward the pole in one part of the cell and the other set moves toward the pole in the other part of the cell.

5. During *telophase* the cell becomes constricted in the middle and divides into two offspring cells. This happens such that each offspring cell receives one of the two sets of chromosomes that were separated during anaphase.

Furthermore, the spindle apparatus now disappears (though a centriole pair remains in each offspring cell), the nuclear membranes reassemble around the chromosomes in each of the offspring cells, and the chromos-

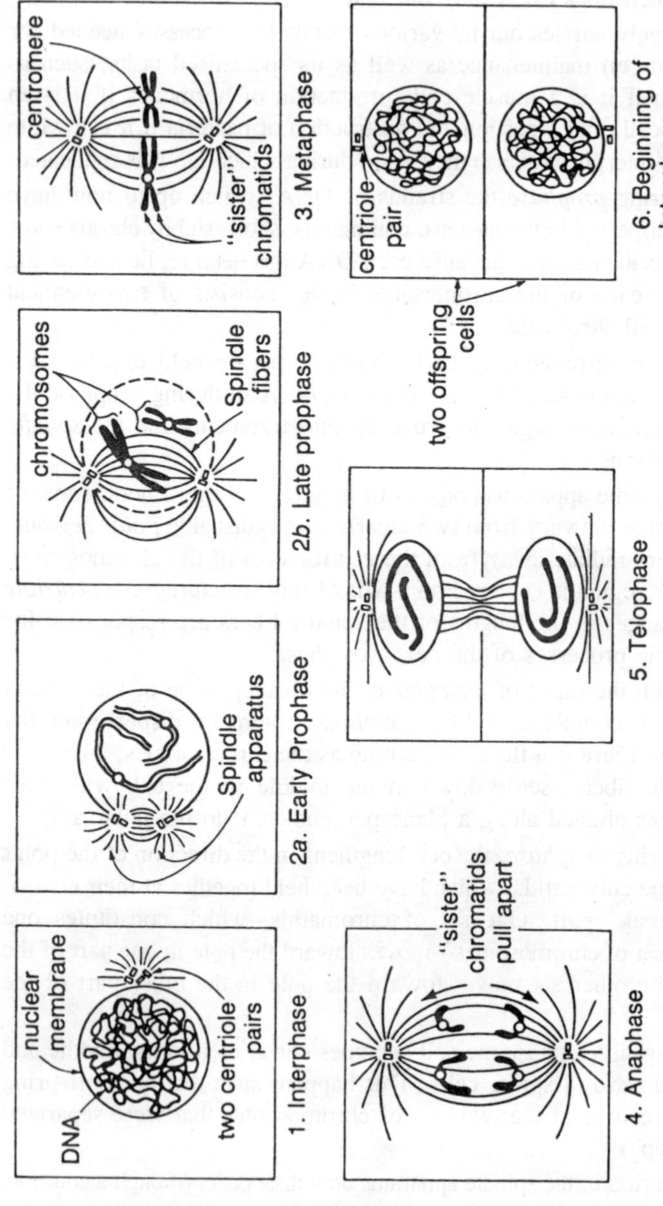

Figure 5.12 : Schematic illustration of the major stages of mitosis in a typical animal cell. 1. Interphase: 2a. Early prophase: 2b. Late prophase: 3. Metaphase: 4. Anaphase: 5. Telophase: 6. The cell divides into two offspring cells by cytokinesis and a new interphase begins for each of the new cells.

omes begin to unwind and disperse throughout the nucleus. Cell division is completed and the two offspring cells enter interphase.

Meiosis

The sequence of nuclear changes that constitute the basis of sexual reproduction in eukaryotic organisms is called *meiosis.* It is more complicated than mitosis because it is *a nuclear reduction* division that divides the set of chromosomes precisely in half.

Meiosis leads to the formation of *reproductive cells* (also called *gametes)—sperm* in the case of males and *eggs* (or *ova)* in the case of females. For example, each human sperm and egg carries 23 chromosomes, not 46 as do fertilized eggs and our somatic (that is, body) cells.

The halving of the number of chromosomes during meiosis is necessary because when a sperm fuses with an egg during sexual mating, the chromosomes of the sperm and egg are combined. Without the halving, the fertilized egg would have twice as many chromosomes as each of the parental cells. Within just a few generations, this would lead to an intolerably large number of chromosomes.

In order to understand meiosis, it is important that you first become familiar with a few terms and concepts. Human fertilized eggs or somatic cells contain 46 chromosomes, which are called *a diploid set.* Half of the chromosomes of diploid sets—namely 23 in humans—come from one parent and the other half come from the other parent.

Each half set of chromosomes is called *a haploid set (haplo-* and *diplo*—are derived from the Greek, meaning "single" and "double," respectively). In general, if the diploid set contains *2n* chromosomes, then the haploid set contains *n* chromosomes (n = 23 in humans, 24 in apes, and 4 in *Drosophila).* During fertilization two haploid cells-an egg and a sperm/1fuse to form a diploid fertilized egg (called *a zygote*).

Each of the 23 chromosomes of a human haploid set is uniquely identifiable, as illustrated in figure elsewhere in this chapter. Cytologists number them 1-22, *plus X* or *Y* for the twenty-third chromosome. Chromosomes with the same identification number are said to be *homologous.* For example, in human diploid cells the two chromosomes numbered 1, which carry genetic information for the same phenotypic traits, are homologous chromosomes, or *homologs* for short.

The same is true of the chromosomes numbered 2, 3, and up to 22. The only exceptions are the X and Y chromosomes. They code for some different traits and are thus only partially homologous. One important

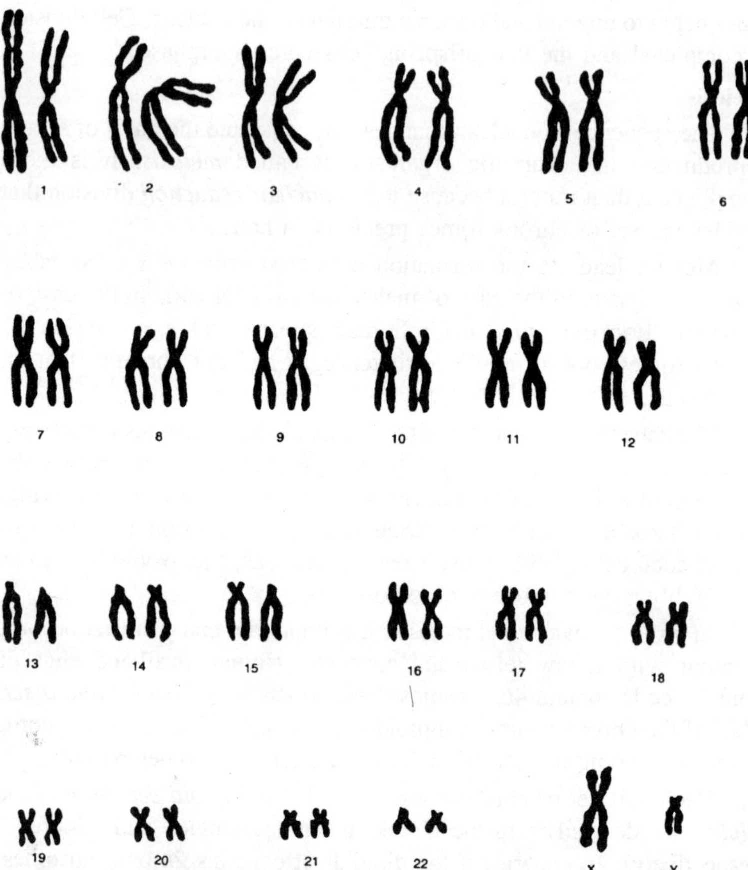

Figure 5.13 : The chromosomes of a human somatic cell. Photograph (a) shows the 46 chromosomes—that is, a diploid set. Half of them-23 chromosomes, or a haploid set—are inherited from the mother, the other half are inherited from the father. In photograph (b) the 46 chromosomes are arranged in homologous pairs, lined up in order of size, and numbered 1-22, plus X and Y for the twenty-third pair. Because the twenty-third pair consists of the combination XY, the chromosomes shown here come from a male.

characteristic they code for is the gender of an offspring: Females possess the combination *XX*, males the combination *XY*.

Meiotic cell division takes place in special sex organs, for example the *testes* in male animals and the *ovaries* in female animals. The purpose of meiosis is to produce eggs and sperm containing haploid sets of chromosomes. This is accomplished by two cellular divisions carried out in sequence. The same phases and many of the mechanical processes that characterize mitosis can be identified in these divisions.

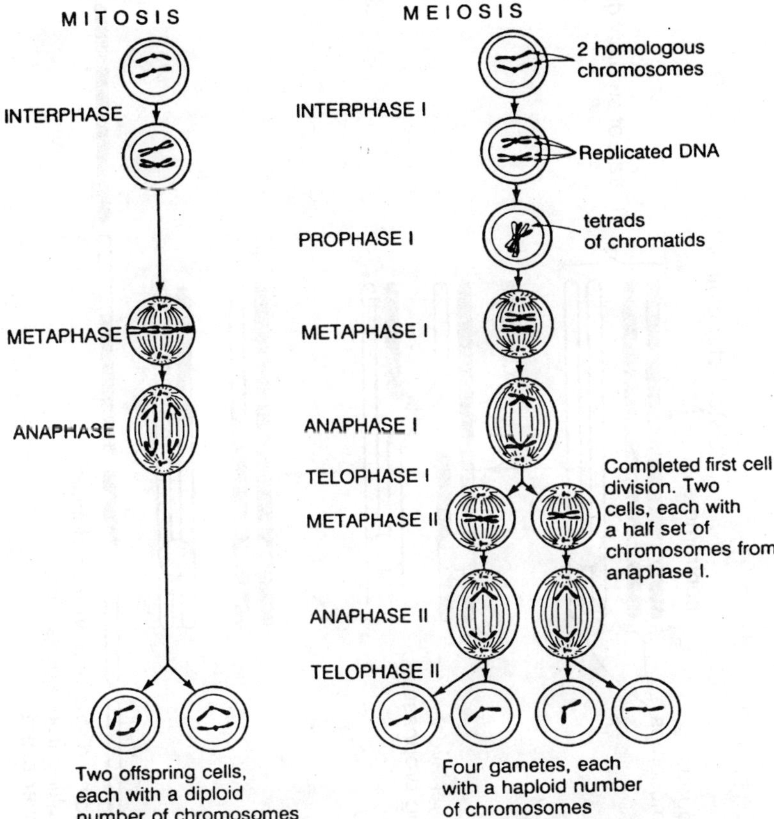

Figure 5.14 : Comparison of mitosis and meiosis. For clarity, the starting cell in each case contains only one set of homologous chromosomes. Note that mitosis consists of one cell division and produces two offspring cells, each with a diploid number (2 in this example) of chromosomes. Meiosis consists of two cell divisions and potentially produces four offspring cells (gametes), each with a haploid number (1 in this example) of chromosomes.

1. *Interphase I* is characterised by the replication of DNA. The two copies of each chromosome remain joined as sister chromatids and behave as a unit throughout the first division.

2. During *prophase I* the strands of DNA coil into short stubby chromosome bodies (each with two sister chromatids), and homologous chromosomes pair up by a process called *synapsis* to form *tetrads of chromatids*.

During this pairing, which persists until anaphase I, pieces of DNA may be randomly exchanged between homologous chromatids. This

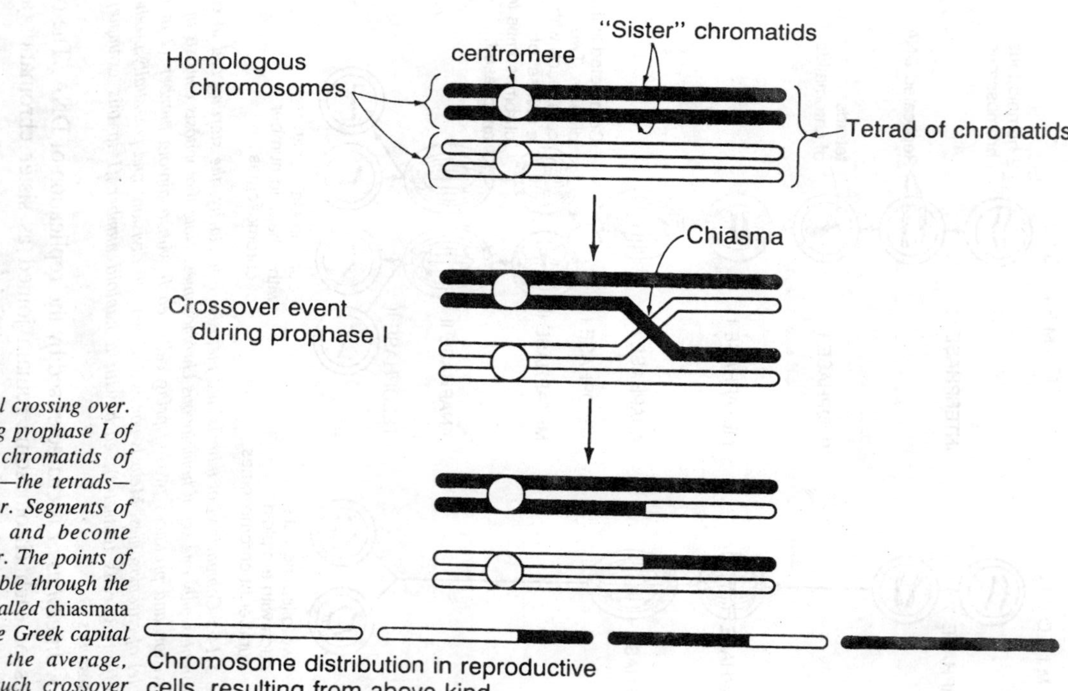

Figure 5.15 : Chromosomal crossing over. This process occurs during prophase I of meiosis, when the four chromatids of homologous chromosomes—the tetrads—are tightly packed together. Segments of chromatids may break and become exchanged, or crossed over. The points of crossing over are often visible through the light microscope and are called chiasmata *(singular* chiasma, *from the Greek capital letter* X *or "chi"). On the average, between two and three such crossover events occur on each pair of human chromosomes during meiosis.*

crossing over, illustrated in figures elsewhere in this chpater, has important consequences for evolution. It scrambles maternal and paternal genes and contributes greatly to genetic variation in sexually reproducing eukaryotic populations.

This scrambling occurs in addition to the random separation of maternally and paternally derived chromosomes during cell division of anaphase I. Also, during prophase I, the nuclear membrane begins to break up and a spindle apparatus forms.

3, 4. During *metaphase I* and *anaphase I,* the tetrads line up along a plane and the spindle fibers pull the homologous chromosomes (each still consisting of two sister chromatids) apart. One of the resulting haploid sets of chromosomes migrates toward one part of the cell, the other set toward the opposite part. This is quite different from mitosis, which separates the sister chromatids of all the chromosomes and sends them to opposite sides of the cell.

The separation of homologous chromosomes during anaphase I of meiosis is completely random and is the major source of variation among sexually reproducing eukaryotes. It is impossible to predict for any of the tetrads toward which side of the cell the paternally and maternally derived chromosomes will travel, except that they will travel in opposite directions. Consequently, each of the separated haploid sets of chromosomes ends up with a random assortment of maternally and paternally derived chromosomes.

5. During *telopbase I* the cell divides by cytokinesis into two offspring cells, with each receiving one of the half sets of chromosomes separated during anaphase I. Nuclear membranes form around the chromosomes and the first division is completed. The two offspring cells now enter the second division.

6, 7, 8, 9. A transient *interphase II,* during which there is no DNA replication, is quickly followed by *prophase II, metaphase H,* and *anaphase II.* Again, these phases closely resemble those of mitosis, except that there are only half as many chromosomes because their number was reduced during the first division. In particular, during anaphase II, the sister chromatids separate and are pulled toward opposite sides of the cells.

10. During *telopbase II,* each of the two cells produced during the first division divides a second time to form four sex cells, each containing a haploid set of chromosomes. Nuclear membranes reassemble and meiosis is completed. Due to the crossing over of pieces of DNA during prophase I and the random separation of homologous chromosomes during

anaphase I, the four gametes produced by meiosis may inherit any combination of maternally and paternally derived genes from the original diploid cell.

Please note that meiosis is only the initial stage in the making of mature gametes. For example, cells developing into sperm need to grow *tails* (flagella), with which they swim, and *acrosomal vesicles,* which contain molecules for aid in penetrating an egg during fertilization.

Eggs need to accumulate yolk for the nourishment of the embryo. Pollen grains (which contain sperm in the flowering plants) need to acquire durable covers and generate a pollen tube, through which the sperm pass to the egg for fertilization.

In human males, all four haploid cells produced by meiosis develop into viable sperm. The process of meiosis goes on at such a high rate that a healthy adult human male can make 100,000 sperm per minute—about 300 million per ejaculation—and he can keep that up throughout much of his life. In contrast, in human females only one mature egg results from each meiotic division.

These divisions begin when the female is still an embryo, but they are for the time being arrested at the first division of meiosis. Only when she reaches sexual maturity, approximately at age thirteen, do the meiotic divisions resume and produce mature eggs—one per menstrual cycle (occasionally more). Altogether, about 450 eggs mature in the course of her reproductive life.

THE BENEFITS OF SEXUAL REPRODUCTION

At a four-day symposium entitled *The Origin and Evolution of Sex,* held during the summer of 1984 at the Marine Biological Laboratory in Woods Hole, Massachusetts, the participants agreed to define sexual reproduction as the process whereby a cell containing a new combination of genes is produced from two genetically different parent cells.

Clearly, this definition applies to eukaryotes, for in this case the genes of the zygote are derived from two genetically different parent cells, namely from a maternal and paternal gamete.

Prokaryotic Sex

Though prokaryotes reproduce by binary cell division and sometimes by budding or spore formation, in certain instances the above- definition of sexual reproduction applies to them as well. The reason is that occasionally bacteria receive segments of DNA from other bacteria. In one mechanism, called *transduction,* a virus carries a bit of DNA from the cell it has grown in to a new cell, but this probably happens quite irregularly. A more important mechanism of gene transfer among bacteria is a process called *conjugation.*

During conjugation, the membranes of two neighboring bacteria fuse at some point, a channel forms between the two cells, and a segment of DNA is transferred from one of the bacteria to the other.

Conjugation resembles eukaryotic sexual reproduction. Quite commonly, it is induced by genetic elements known as *plasmids*. The plasmids bring about conjugation by forming proteins on the bacterial surface that make the cell sticky for other cells.

This, in turn, induces the fusion of the cell to a neighboring cell and the formation of a channel between them. The plasmids of the first cell then make copies of themselves, which are transferred to the second cell. Quite often the plasmids promote the transfer of bacterial DNA as well, in addition to their own DNA.

The plasmid-carrying cell (that is, the one that induced conjugation) thus becomes a "male" or gene donor, while the second cell becomes a "female" or gene recipient. After conjugation, the two bacteria break apart and go on to grow as before, though sometimes the donor cell dies because it has given away some of its genes.

After the transfer of a segment of DNA from one bacterium to another has taken place, part of the segment is sometimes spliced into the circular DNA of the recipient bacterium and thereby becomes part of its genetic makeup. This splicing is called *recombination* and requires a host of enzymes to orchestrate the breaking, inserting, and fusing of the DNA strands.

Many of these enzymes are the same as those used in the repair of UV damaged DNA, suggesting that prokaryotic sex was able to arise because part of the DNA repair machinery had already evolved earlier. Note that neither conjugation nor transduction entail cell reproduction, as does eukaryotic sex. Rather, these processes merely refer to transfer of genes from one bacterium to another.

Unlike eukaryotic sexual reproduction, recombination in bacteria is a rather imprecise process. Rarely does exactly half of the recombinant DNA come from one parent cell and the other half from the other parent cell.

For example, in some cases the transferred segment of DNA carries just one or a few genes, and in others it carries nearly the entire set of the donor's genes. And in many instances, it involves only plasmids and not bacterial DNA.

Despite its imprecision, sexual reproduction has a profound effect on the evolution of prokaryotic populations. By shuffling and spreading genes, it creates an enormous degree of variation, upon which natural selection can act.

Beneficial genes, which impart advantages on their carriers in the struggle for survival and hence have an above-average chance of being passed on to future generations, tend to spread and accumulate in a population; detrimental genes, which impart disadvantages, tend to diminish.

The result is that bacterial populations are able to adapt remarkably rapidly to changing conditions in the environment. For example, during just the past few decades, conjugation and plasmid transfer have spread resistance to antibiotics widely among pathogenic bacterial species. Other bacterial species have acquired the ability to make bacteriocidal chemicals.

And still others have evolved the ability to break down unusual organic compounds, such as those present in oil spills. Such rapid adaptations are not possible in the absence of sex; for in these cases adaptation depends on the occurrence of favourable mutations, which are relatively infrequent in a single organism.

Eukaryotic Sex

Given the adaptive advantages of sexual reproduction among prokaryotes, it is not surprising that it also evolved among eukaryotes. In fact, eukaryotes may have evolved from prokaryotes that already had sex, such as gene transfer by conjugation, though no one knows for certain. However, we do know that eukaryotic sex became much more complex and refined than prokaryotic sex.

This probably happened in part because eukaryotic DNA is distributed over two or more separate chromosomes. Any process less precise in distributing chromosomes than meiosis followed by fertilization would mean that some offspring cells might end up with extra genes or chromosomes, while others might end up with too few. Such processes would rapidly degrade genetic information and lower the viability of eukaryotic organisms.

Despite the apparently plausible argument that sex evolved because of its adaptive advantages -namely the enormous degree of genetic variation it creates-some biologists have recently begun to question this argument. For example, Norton Zinder, a molecular geneticist at the Rockefeller University in New York, explains the problem this way: "How could an organism that only passed half of its genes to its offspring [through sexual reproduction] ever have competed with [an asexual] progenitor that passed all of them? It seems unlikely that the offspring produced sexually were `fitter' than their asexually produced relatives".

In the face of questions such as this one by respected scientists about the value of sex, the participants at the Woods Hole symposium on *The Origin and Evolution of Sex* agreed that at present no one really understands why sex persists. With this sentiment, they reflected the thoughts Charles Darwin expressed on this topic in 1862: "We do not even in the least know the final cause of sexuality. The whole subject is as yet hidden in darkness".

It is probably fair to say that today the subject is less hidden than it was during Darwin's time, before Mendelian genetics and the structure of DNA became known, but uncertainties do remain and the search to understand the full benefits of sexual reproduction continues.

In spite of these uncertainties, it is clear that with the origin of eukaryotic cells the pace of biological evolution began to quicken. The new cells were larger than their prokaryotic predecessors and contemporaries. They carried out their aerobic metabolisms with the aid of mitochondria and chloroplasts (if they were photosynthesizers), which increased their ability to generate energy quickly.

They possessed more DNA and, hence, genetic information. And they were able to execute more complex functions, such as mitosis and meiosis. The combination of these abilities, including the ability to reproduce sexually through meiosis, was probably responsible for the increased pace of the evolution of eukaryotes.

In any case, only eukaryotes have evolved into complex macroscopic and multicellular organisms such as plants and animals. This evolution was already well under way by roughly 670 million years ago and it has continued to the present. It has created the enormous variety of organisms we see all around us today.

6

Solid and Water Contaminants

We began this book with an examination of major contamination of soil and water in Niagara Falls, New York, at the Love Canal disposal site. Contamination results from the introduction of potentially harmful chemical and biological substances into hydrologic and soil-water systems. Love Canal was presented as a symbol of the hundreds, or even thousands, of instances of contamination being reported throughout the world by government agencies, private study groups, and academic analysts.

An examination of indexes to major newspapers in North America and Europe demonstrates that soil and water contamination has been the object of steadily rising concern in the last third of the 20th century. This contamination will become a major preoccupation of analysts during the last decade of the century.

The subject of soil and water contamination encompasses much of what we know about soil and water resources, the biological effects of contaminants, and the changes in ecosystems affected by contaminants. It is a subject that involves geologists, biologists, and environmental engineers.

Here the subject is viewed from the geological perspective, specifically from the hydrogeologist's perspective, since much of the research on which this chapter is based comes from the hydrogeologic literature. We begin by reviewing the work done by "environmental protection" agencies in dealing with soil and water contamination in many different countries and in smaller political units. The Love Canal case required the U.S. Environmental Protection Agency and the New York Department of Environmental Conservation to integrate results from hydrogeological, geochemical, and epidemiological investigations.

Thus those who deal with contamination hazards find it necessary to become familiar with findings from the fields of ecology, toxicology, and epidemiology, as well as hydrogeology and engineering.

Given the breadth of the topic, this chapter will touch on some quite disparate subjects. First, we will look at the kinds of contaminants with which we are concerned and some of the reasons for this concern. An effort is made to list the sources from which these contaminants are known to originate.

This will establish the need for hydrogeologic studies directed at understanding the processes by which contaminants find their way into soil and water and then reach their targets. These processes are illustrated by considering a few well—documented cases in which hydrogeologists have been able to establish the pathways that the contaminants have followed, the transformations that they have undergone while moving along these pathways, and the barriers that they have encountered.

These examples also illustrate the kinds of monitoring systems that are employed and the kinds of investigations that have been made of soil and water contamination. In some instances, these studies also show the harm done.

With all of this technical material as background, we will then be in a position to examine tactics for dealing with contamination problems-tactics that include technological approaches, legislated preventive measures, and legislated systems for compensating those affected adversely by exposure or by the possibility of exposure.

CONTAMINANTS

Soil and water may be contaminated by airborne and waterborne natural or manmade substances. These substances are introduced directly by human actions such as the application of fertilizer and pesticides, and through the disposal of wastes. The number and variety of contaminants are very large, so we will begin by establishing some major categories of contaminants.

In discussing a specific contaminant we will indicate the category to which it belongs, since a problem associated with a specific contaminant may be seen as symbolic of the problems associated with the entire category.

Each of the substances listed in this section becomes a matter of concern only when it appears in a setting in which it may cause harm or may affect the functioning and character of an ecosystem. In order to cause harm, the substance must find a pathway to a target that for some reason is vulnerable to attack. The contaminant must be present at a concentration above that of the background level, since organisms and ecosystems are adapted to substances present at these levels.

Once a contaminant reaches a target, what are the adverse effects that may result? The most readily apparent are acute health effects, which manifest themselves immediately upon exposure. In humans, these might include nausea; eye, lung, or skin irritation; and dizziness. In animal populations, acute health effects may be represented by sudden mortality, as might be seen in fish in a stream that has suddenly become contaminated.

Acute health effects are far less common than other delayed health effects that develop and affect their targets over substantial periods of time. These long—lasting effects are called *chronic health effects*. They are represented by a variety of diseases and by progressive malfunctions of the liver, kidney, lungs, and central nervous system, by cancer and genetic damage, and by reproductive system malfunction.

These delayed and long-lasting effects are far more serious, and they are far more difficult to identify, because many of them are already represented in the target population under normal conditions.

Major categories of contaminants in common use are as follows:

1. *Pathogenic microorganisms:* These have their origins primarily in human and animal wastes. They include bacteria and viruses that cause diseases such as hepatitis, encephalitis, and typhoid fever, and that can also cause a variety of gastrointestinal problems. For example, hepatitis outbreaks have been triggered in the United States by the consumption of shellfish contaminated by sewage.

   The standard indicators of contamination by microorganisms originating in fecal wastes are the coliform bacteria represented by *E. coli* and similar organisms normally present in fecal material. These coliform bacteria are not in themselves pathogenic, but they are evidence of potentially dangerous contamination. Coliform organisms of nonfecal origin may also be observed. These commonly indicate earlier contamination by fecal material or other nonfecal sources.

2. *Inorganic chemicals:* Inorganic compounds and their components

are normally present in soil or water that has not been contaminated. Therefore it is always necessary to establish the natural or *background* concentrations of these substances before attempting to evaluate the degree to which soil or water is contaminated by them.

Inorganic contaminants (exclusive of radioactive nuclides) include the following: the chloride ion Cl- ; heavy metals including lead (Pb), mercury (Hg), cadmium (Cd), chromium (Cr), and nickel (Ni); the nutrients phosphorus (commonly present as phosphate ion PO, = and nitrogen commonly present as nitrate ion NO_3- or nitrite ion NO_2- ; and sulfur (commonly present as sulfate ion SO_{4-}). The sum of all dissolved substances is referred to as TDS (for total dissolved solids). Inorganic chemicals, especially the heavy metals, can be toxic. Mercury becomes even more toxic after being converted by microorganisms to the form known as *methyl mercury*. Nitrite present in ground water is known to have caused toxic effects in cattle and is responsible for *methemoglobinemia* (blue baby disease) in infants.

Loading of lakes with the nutrients phosphorus and nitrogen is also important because it stimulates the production of particular selected species of algae, leading to a decline in water quality and species variety in these lakes. Loading of lakes, streams, and soils with acids causes major changes in the associated ecosystems.

3. *Organic chemicals:* These include thousands of synthetic chemicals used in industrial processes, pesticides, food additives, and drugs. Some of these of special concern during the period 1960-1984 are: *low-molecular weight chlorinated hydrocarbons,* including trichlorethylene, carbon tetrachloride, tetrachlorethylene, 1,2-dichlorethane, and vinyl chloride; and *pesticides* such as toxaphene, endrin, methoxychlor, lindane, 2,4-D, 2,4,5-T, and DDT. Other organic contaminants of concern include *polybrominated biphenyls (PBBs), polychlorinated biphenyls (PCBs), dioxin,* and *benzene.*

   Low-molecular weight hydrocarbons such as trichlorethylene and vinyl chloride are known carcinogens that cause cancer in laboratory animals or humans, even at rather low concentrations. There may be no safe lower threshold of exposure to these substances, and the Environmental Protection Agency has issued guidelines for states and municipalities to employ in evaluating risks from exposure to these substances.

Pesticides such as 2,4,5-T have been implicated by animal tests and human evidence as having the potential for causing miscarriages, birth defects, and other adverse reproductive effects. In some instances, these adverse effects are caused by the contaminant 2,3,7,8-tetrachlorodibenzo-p-dioxin, also known as TCDD and dioxin.

The Environmental Protection Agency indicated concern about these and other pesticides when it suspended registration for them on 28 February 1979. At that time, the only use not suspended was use on rice fields in Arkansas. Since then, the medical profession has expressed concern about possible chronic health effects in regions where 2,4,5-T continues to be used.

4. *Radioactive nuclides*. Unstable forms of the chemical elements are called *radioactive isotopes* or *radioactive nuclides*. Each of these breaks down or *decays,* until a stable, nonradioactive isotope is formed. For example, carbon can occur as the radioactive isotope called *carbon 14,* or in chemical nomenclature, ^{14}C. Carbon 14 decays to nitrogen 14, giving off an electron (beta particle) in the process.

   Each radioactive nuclide may be characterised in terms of its half-life, the time required for one-half of an initial mass of the nuclide to decay. The half-life of carbon 14 is 5730 years, which means that any given mass of carbon 14 present at a particular time will contain only half as much 5730 years later. The half-life commonly is indicated by a time unit placed before the chemical symbol-5730 yrs ^{14}C.

   Radioactive nuclides have their origin in naturally radioactive materials, in commercial and military nuclear-fission reactors, in nuclear weapons testing, and in other less important sources. They include the following:

Tritium ^{3}H	Strontium ^{90}Sr
Cesium ^{137}Cs	Iodine ‘^{31}1
Radon ^{222}Rn	Plutonium ^{239}Pu
Krypton ^{115}Kr	Uranium 235U Uranium 238U

   Each radioactive nuclide can emit *ionizing radiation* as it decays to other radioactive or nonradioactive nuclides. This ionizing radiation, which may be in the form of charged alpha and beta particles or in the form of neutral gamma rays, causes atoms and molecules to be electrically charged, or ionized.

The harmful effects of this radiation arise because it damages the deoxyribonucleic acid (DNA) located within the genes that control the normal functioning of all living cells. This damage may be caused by ionization of material in the gene, or of the molecules adjacent to the DNA, which then react with the DNA.

The radiation damage ultimately takes the form of a mutation in the living cells. The summed effects of these mutations take three forms: (i) acute somatic (nongenetic) damage manifested as radiation sickness and even death; (ii) delayed somatic damage manifested as cancer, birth abnormalities, and a shortened life span; and (iii) genetic damage caused by nonlethal effects on reproductive cells.

This leads to an increase in birth defects and to genetic effects passed on beyond the generation to which the initial damage was done.

5. *Particulate material:* Some contaminants consist of identifiable particles of natural minerals. These are represented by the following: (i) soil particles including clay minerals; and (ii) mineral fibers such as asbestos.

   These particles may be harmful or they may carry harmful elements with them. For example, some forms of asbestos are known carcinogens, implicated especially in causing lung cancer. Clay minerals can adsorb radioactive elements such as cesium 137, making them carriers of radioactivity.

ASSESSMENT OF HARM

Assessing harm-whether to individuals, populations, or ecosystems-is a difficult scientific task. Predicting future harm or evaluating risk (see Chapter 4) is even more difficult. In the field of human health, assessing harm is the task of the science of *epidemiology,* which studies human populations to establish anomalous patterns of disease or death.

Epidemiology also tries to find explanations for these patterns by studying patterns of exposure to contaminants. The record on the assessment of health effects at Love Canal and other chemical disposal sites shows how difficult this is. This record goes far beyond what can be covered in this book, but we can list some situations in which contamination has either caused harm or where the potential for harm has called for action before the harm was fully proved. It is not only direct harm to human health that concerns us.

Contaminants can also harm individual plant or animal species, and this can lead to changes in the entire ecosystem in which these species are found. An ecosystem is a stable assemblage of plant and animal species and the physical surroundings in which they exist.

Wilson in a study of critical environmental problems, found that the most significant aspect of human action was not the harm caused by a single specific pollutant but rather the total impact on an ecological system that arises from the sum of all introduced contaminants.

An ecosystem may be affected at many levels with an effect at any one level having the potential to affect other levels adversely. For example, a predator such as a bird species might be directly affected by a contaminant that had been ingested.

The species might also be affected indirectly by the loss of its food supply due to the effect of contamination on the organisms making up that food supply. Such a species might also be affected by damage caused at a particular stage in the reproductive cycle. Here we summarize findings reported for DDT by Wilson and by Garrels, Mckenzie, and Hunt.

The pesticide DDT and its metabolite DDD are known to affect the growth, reproduction, and mortality of animal species including plankton, fish, crustacea, mollusks, and birds. Fish, for example, begin to suffer reproductive failure when DDT concentrations in the fish reach a level of 5 to 10 ppm.

Thus they may be affected directly as a result to uptake of DDT. In addition, they are capable of concentrating DDT by the process of *bioaccumulation, so* that they display higher DDT concentrations than are displayed in their food supply.

Since these fish serve in turn as food supplies for such bird species as the bald eagle, the osprey, and the brown pelican, DDT is accumulated in all of these bird species. The DDT reduces hatching success of these birds and prevents the formation of shells of normal thickness.

As a result, where these bird species feed on fish supplies contaminated with DDT, these species may be eliminated, at least locally. This phenomenon accounts for the decline of the bald eagle, the osprey, and the brown pelican in contaminated areas.

Table 6.1 : Contamination Incidents.

Location	*Nature of Contamination*	*Harmful effects observed or Predicted*
Sweden and Norway	Precipitation containing high concentrations of sulfate ion making the precipitation more acid than normal.	Elimination of fish from acidified lakes, changes in the lake ecosystem, release of metals from soils to ground and surface water, possible adverse effects on human health and forest productivity (Swedish Ministry of Agriculture 1982; Committee on the Atmosphere and the Biosphere, 1981).
Colorado	Contamination of ground water with chemical wastes released by Rocky Mountain Arsenal.	Death of sheep, changes in plant viability and chemistry, and contamination of ground water making it unfit for use.
Missouri	Contamination of soil with dioxin (1971-1983).	Death of animals, evacuation of Times Beach, Missouri, and potential evacuation of other contaminated localities. Possible human health effects under investigation.
New York State	Contamination of soil and ground water with industrial chemicals.	Evacuation and abandonment of Love Canal and extensive remedial engineering at Hyde Park-both in Niagara Falls, New York. Possible health effects under examination.

SOURCES OF CONTAMINATION

There are hundreds of different sources from which contaminants may enter soil and water. Each source has a specific set of contaminants, and each is coupled to the hydrologic system in a specific way. We will look at the kinds of sources commonly encountered and then at the process of contamination in various settings.

One of the most comprehensive and readily available documents dealing with sources of contamination is the U.S. Environmental Protection Agency's *Report to* the Congress on Waste Disposal Practices and Their Effects on Ground Water. This report provides a framework within which most terrestrial sources are examined here.

In addition, we will also deal with a number of sources for airborne contamination.

Terrestrial sources of contamination are divided into two large groups: *point sources* and *nonpoint sources.* Point sources of contamination are those that would appear to be isolated points when plotted on a map.

The distinction between point and nonpoint sources is made because point sources are more easily identified and controlled than nonpoint. Defining a source as a point source depends on the scale of the maps we are using. Point sources include:

1. Septic tanks and cesspools (domestic wastes).
2. Collection and treatment systems employed in handling municipal wastes (domestic and industrial wastes).
3. Land on which sludges or effluents from sewage-treatment systems are spread (residues from municipal sewage-treatment systsm).
4. Animal feed lots (animal wastes).
5. Landfills and dumps (domestic and industrial wastes exclusive of sewage).
6. Industrial waste impoundments (industrial wastes).
7. Mines (mine wastes consisting of inorganic chemicals and minerals).
8. Drilling operations (brines or saline waters brought to the surface).

Nonpoint sources of contaminants are those that would appear as sizable areas rather than as isolated points on a map. Major nonpoint sources are listed below:

1. Agricultural land and even suburban grassed areas (fertilizers and pesticides).
2. Irrigated agricultural lands (salts contained in the irrigation water).
3. Urban areas (oil, gasoline, spilled materials).
4. Air masses (airborne contaminants).

Point and nonpoint sources may be represented in a single schematic cross section.

Sources of atmospheric contamination include fossil-fuel power plants, nuclear power plants, mineral-processing smelters, chemical plants, and all internalcombustion engines. Individual stacks at these plants are point sources as are individual internal-combustion engines or wood-burning stoves. Because engines and stoves commonly occur in

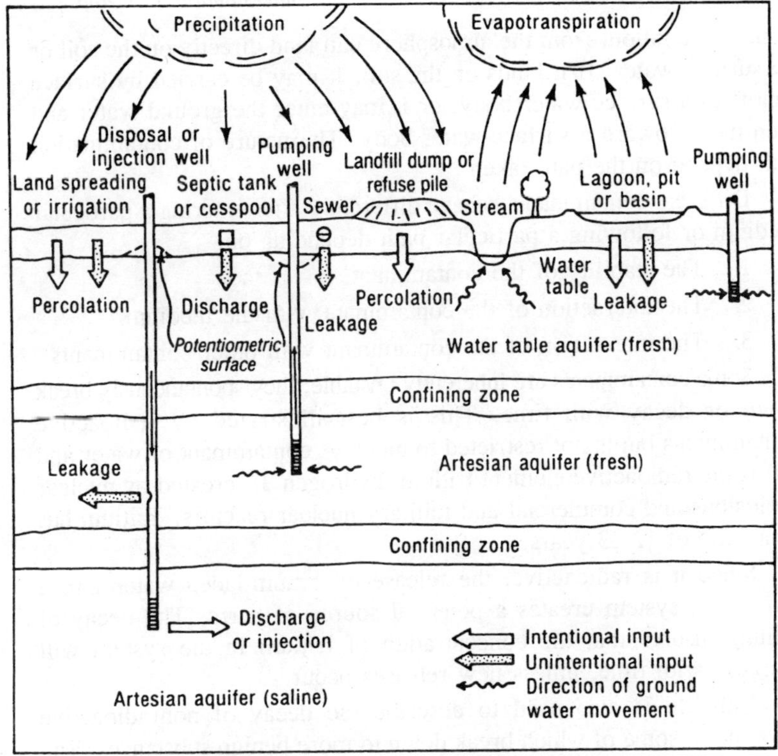

Figure 6.1 : Schematic cross section showing sources of ground-water contamination.

large numbers over large areas, they collectively create major nonpoint sources of atmospheric contamination.

PROCESSSES OF CONTAMINATION

General Considerations

Soil and water contamination takes place gradually, so we view these phenomena as processes that we can monitor and describe in terms of various stages. These processes depend on the nature of the source, the characteristics of the medium being contaminated, and the nature of each contaminant moving through that medium.

These processes are often complicated, but we can identify some principles that govern each stage of the contamination process. A contaminant can enter the soil, surface water, or ground water directly or by first passing through other elements of the system. For example, radioactive fallout from the atmosphere can land directly on the soil or on surface water. If it lands on the soil, it may be carried by surface runoff to a surface-water body, or it may enter the ground water and then move toward a surface-water body. The nature of contamination will depend on the path taken.

Different contaminants behave differently after entering a particular medium or following a particular path depending on:

1. The stability of the contaminant.
2. The interaction of the contaminant with the medium.
3. The interaction of the contaminant with other contaminants.

Some contaminants are inherently unstable; they spontaneously break down or decay with time. This is best illustrated by radioactive contaminants but is not restricted to them. A contaminant of water and soil is the radioactive element tritium (hydrogen 3), created by nuclear explosions and commercial and military nuclear reactors. Tritium has a half-life of 12.33 years.

Since it is radioactive, the release of tritium-laden water into a hydrologic system creates a potential source of harm. The decay of tritium ensures that the concentration of tritium in the system will decrease with time, unless new releases occur.

Half-life is also used to describe the decay of nonradioactive substances, some of which break down to more benign substances. The pesticide DDT gradually breaks down to other compounds, some toxic and some not. The half-life of DDT is estimated to be between 10 and 20 years.

Although the half-life of an organic compound cannot be specified in terms of a single precise value, knowledge of an approximate half-life is useful in estimating the time required for an organic contaminant to reach low levels of concentration.

We will be especially concerned here about any harmful substance that has a relatively long half-life or that we describe as *persistent.*

Contamination from Atmospheric Sources

The contamination of soil and water by deposition of contaminants from atmospheric sources is important. The term *fallout* is used to describe the deposition of radioactive contaminants from the atmosphere. The phenomenon is well represented by studies of radioactive contamination, dioxin and contamination in Seveso, Italy, and acid precipitation. In every instance, a contaminant is released from a source, transported through the air, and deposited on land or water. Here we will examine

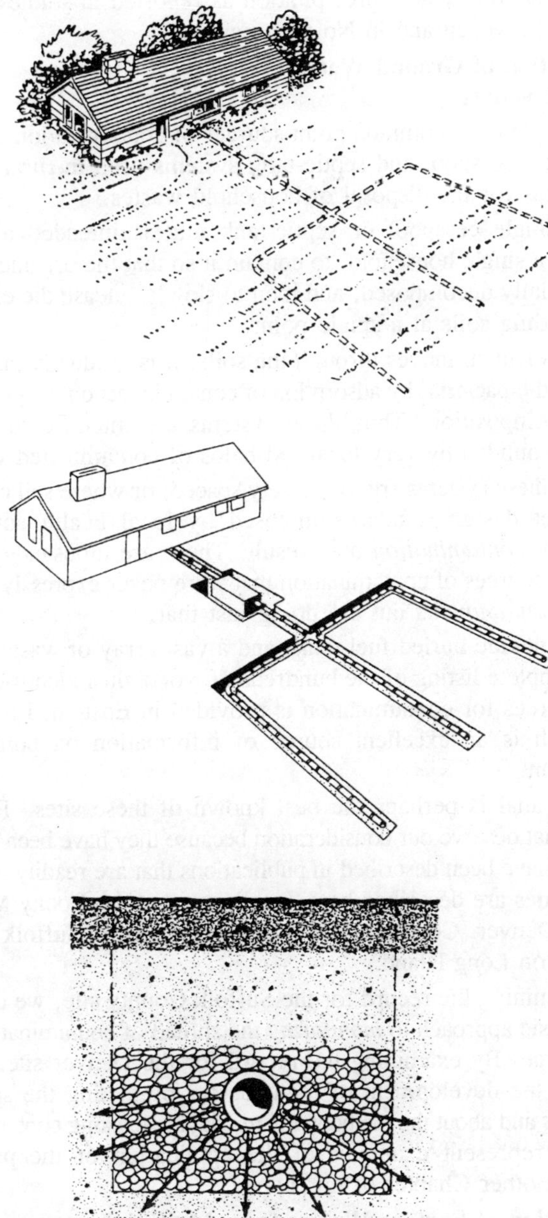

Figure 6.2 : Septic system leach field showing detail of stone-filled trench.

the contamination by acid precipitation as reported in studies made in Norway and Sweden and in North America.

Contamination of Ground Water from Point Sources

The single most common point sources of contamination of ground water are the cesspool and septic-tank systems used in rural and not-so-rural areas for the disposal of household wastes.

Each single cesspool or septic-tank unit is intended to receive waste from a single household, to contain it so that the organic material will be partially decomposed, and then to slowly release the effluent in the surrounding soils at a single point.

As the effluent moves through the soils, it is gradually purified by filtration of the bacteria, by adsorption of contaminants onto soil particles, and by decomposition. Thus these systems are intended to be point sources surrounded by very localised halos of contaminated water.

Where these systems are too closely spaced, or where soil conditions do not meet design standards imposed by local health authorities, unacceptable *contamination* may result. There are thousands of much larger point sources of contamination that were never expressly intended to release *contaminants* but are doing just that.

These include buried fuel tanks and a vast array of waste disposal sites. A complete listing of the hundreds of worst sites identified by the EPA as sources for contamination is provided in Epstein, Brown, and Pope, which is an excellent source of information on point-source contamination.

Love Canal is perhaps the best known of these sites. There are other sites that deserve our consideration because they have been carefully studied and have been described in publications that are readily available. Two such sites are described here, one located at the Rocky Mountain Arsenal in Denver, Colorado, and the other located in Suffolk County, New York, on Long Island.

By examining the record for the Suffolk County site, we can learn about the basic approach to *monitoring* the spread of contamination from a point source. By examining the record for the Denver site, we can learn about the development of methods for predicting the spread of *contaminants* and about the methods for remedying existing contamination. Both cases represent examples of the application of the principles described in other Chapter of this book.

Contamination of Surface Waters from Multiple Sources

Surface waters can be contaminated from the whole range of sources listed earlier in this chapter. Mine workings and mine tailings constitute major point and nonpoint sources of contamination. These commonly cause acidification of surface waters.

This results from the oxidation of sulfide minerals and the resultant formation of sulfuric acid. The contamination is similar to that described for areas affected by acid precipitation. Excellent sources on acid-mine drainage are found in the National Research Council's report on coal mining.

Sewage treatment plants and industrial outfalls are sources of contamination that have caused many serious problems in the past. The most common of these problems is the depletion of dissolved oxygen in the receiving waters, a depletion caused by the anaerobic decomposition of the introduced wastes.

Less common, but quite serious, problems have also resulted from the release of toxic chemicals such as Kepone (James River, Virginia), PCBs (Hudson River, New York), dioxin (Times Beach, Missouri), and mineral particles such as asbestos (Lake Superior). Many of these sources of contamination have been brought under control.

As these controls have been introduced, it has been recognised that other, less easily controlled sources of contamination are contributing to water-quality problems. These new sources are atmospheric, as we have seen for acid rain, and terrestrial.

The effect of multiple or dispersed sources is well illustrated by the problems of lakes, bays, and estuaries in areas subject to human influence. These bodies of water are subjected to intense use and are affected by loading from many contaminant sources.

The problems are obvious in the many small lakes that the U.S. Geological Survey has described as *urban lakes* and *real-estate lakes* in the series on Water in the Urban Environment (see, e.g., Britton, Averett, and Ferreira, 1975; and Rickert and Spieker, 1971). Similar and larger-scale problems are also recognised in lakes as large as the Great Lakes, in bays such as Chesapeake Bay, and in estuaries.

It is convenient to examine small lakes here since they illustrate processes that act on many scales. Any lake may be viewed as a complex ecosystem consisting of a physical framework comprising lake basin, soil, and water, and a biota of plants and animals.

A continuous exchange of chemical elements and compounds takes place among all parts of the ecosystem as shown diagrammatically in Figure elsewhere in this chapter. This exchange is facilitated by mixing of upper lake waters *(epilimnion)* and deeper waters *(hypolimnion)* in the

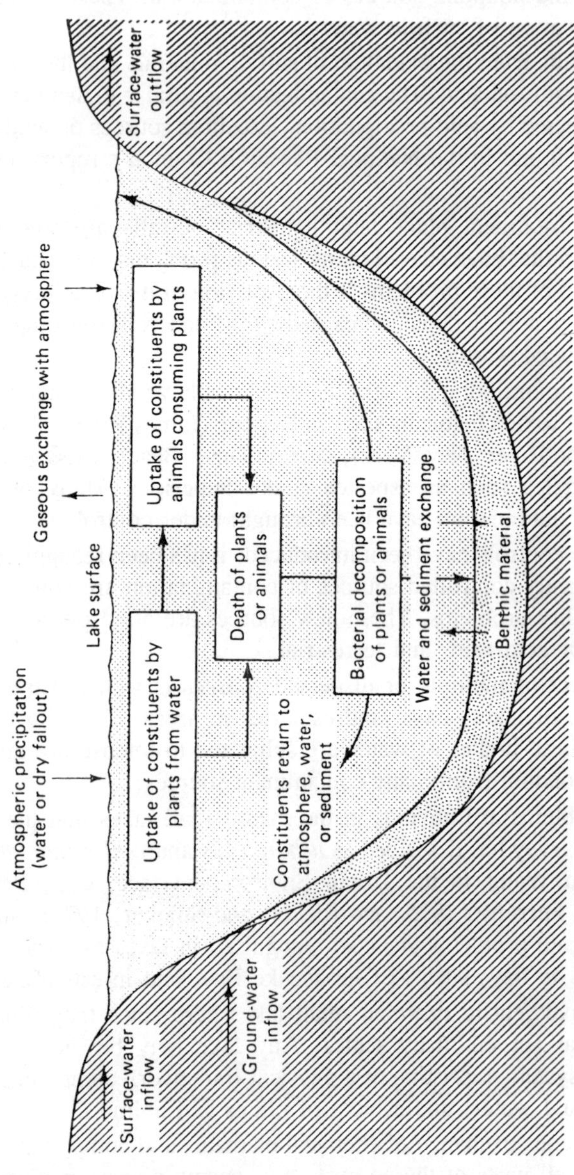

Figure 6.3 : Chemical exchanges among parts of alke ecosystem.

fall and spring when the epilimnion is cooled or heated, causing changes in the density of the surface water. (The terms *epilimnion* and *hypolimnion* are used to describe the stratification of lake waters that results from temperature and density differertces.) When these fall or spring changes occur, the lake waters "turn over" and become mixed.

Under natural conditions, the plant and animal assemblages in a lake display a considerable degree of stability; changes in the character of the lake do take place, but they occur very slowly as seen from a human perspective. Each lake evolves through stages that end with the filling of the lake basin. This evolution ordinarily lasts thousands of years.

In the early stages, lake waters are rich in oxygen, and the concentrations of dissolved solids are relatively low. In such lakes the productivity of plants is limited by the availability of nutrients, and this in turn places limits on the size of animal populations. These lakes are described as *oligotrophic*. As lakes evolve in the natural state, the waters become more nutrient-rich, and a more abundant plant life develops.

At first this increased plant life makes possible a more abundant and varied assemblage of animals as well. As the production of phytoplankton and algae increases, however, a stage may be reached at which the dissolved oxygen content of the lake waters is reduced. This changes the character of the lake drastically. We say that a lake has become *eutrophic* when this stage has been reached.

This evolutionary process may be represented diagrammatically in terms of the production of organic matter per unit area of lake surface. During the oligotrophic stage, the rate of production of organic matter increases slowly.

Then, during the natural eutrophic stage, the rate of production increases rapidly. Finally the rate of production decreases, and the lake fills with sediment and becomes extinct.

Where human influence is significant, the eutrophication process may be accelerated, increasing the rate of production of organic matter very rapidly as shown by the dashed lines in the diagram. As a result, a lake can quickly enter a state that would take thousands of years to reach under natural conditions. Let us examine this process.

Lakes and bays that become surrounded by urban development show a phenomenon called *nutrient loading,* created by inputs of nutrients from point sources such as sewage treatment plants (STPs), and by inputs of nutrient-rich runoff from agricultural and residential land on

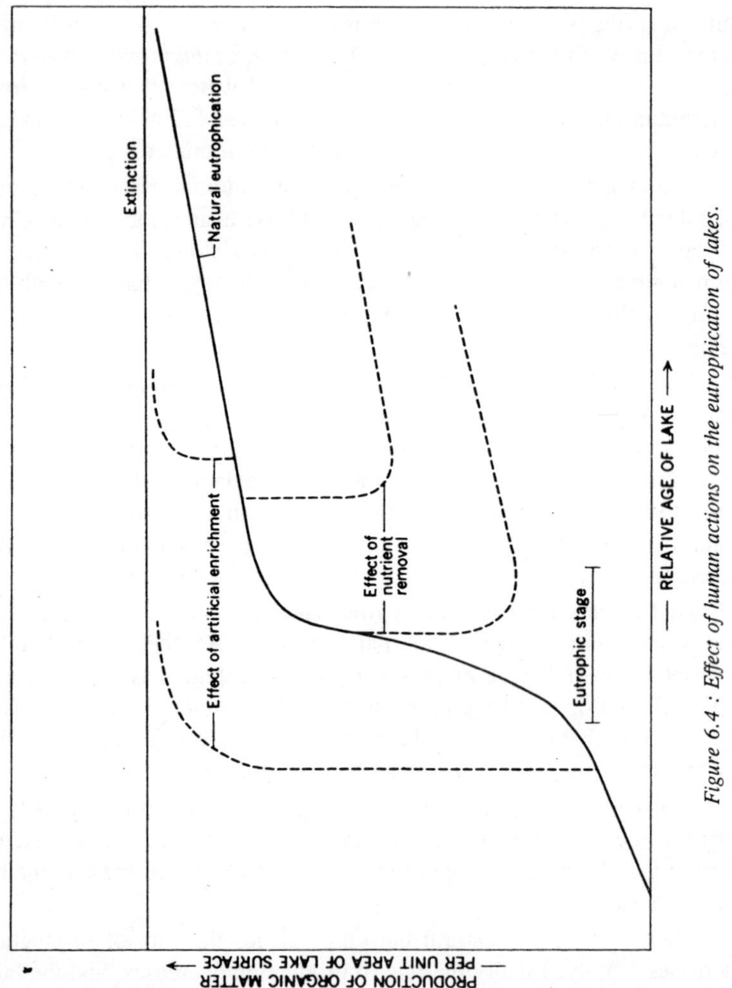

Figure 6.4 : Effect of human actions on the eutrophication of lakes.

which fertilizers have been applied. As a result of this loading, lake waters may become so rich in nutrients that the production of algae is no longer limited by the availability of one or more critical nutrients. When this happens, phytoplankton productivity may cause *algal blooms,* which cover the lake waters with mats of algae.

Algal material in these mats dies and sinks below the surface, decomposing as it sinks and consuming oxygen from the subsurface waters. Then the decomposition products are released to the surface waters. This process drastically changes the character of the lake. Fish kills may occur, and the fish population may also change through a modification of the kinds of species present and through a decrease in the total species variety.

Many urban lakes and reservoirs are also affected by being filled rapidly with sediment eroded from the watershed. The basins of small lakes can be completely filled in periods measured in years. In addition, the incoming sediments carry adsorbed nutrients and a whole array of contaminants.

Once these sediments come to rest, they constitute a source from which nutrients and contaminants can be released over time. This is important because it may affect efforts to restore lakes that have been partially eutrophied and contaminated, as described in the next section on restoration.

Somewhat comparable changes are observed in much larger bodies of water, where such changes take on greater economic importance. This is illustrated by changes taking place in Chesapeake Bay, which is a large estuary fed by a number of rivers draining the eastern seaboard of the United States. These rivers include the Susquehanna and the Potomac, both major rivers draining heavily urbanised watersheds.

Chesapeake Bay is an important commercial fishery and recreational resource. Research showing long-term decreases in oxygen content of bay waters and changes in plant productivity have stimulated much concern. Officer et al. have shown that the volume of Chesapeake Bay waters that becomes *anoxic* (oxygen-free) in the summer has been increasing rather steadily since the 1950s, if not earlier.

This may be seen from maps of Chesapeake Bay for the summers of 1950 and 1980, which show that by the summer of 1980, most of the Chesapeake Bay bottom was anoxic (oxygen-free) or hypoxic (very low dissolved oxygen).

Officer and co-workers argue that these chemical changes in Chesapeake Bay waters apparently are having large-scale ecological effects, causing declines in commercially important benthic organisms such as oysters, and changes in the populations of organisms on which commercial fish species feed. They attribute the anoxia to increased nutrient input into the bay.

The changes in the composition of lake waters and in the character of lake biota are difficult to reverse. Because of this, it is widely agreed that the best way to deal with such problems is through watershed management that controls input before trouble can arise.

However, many lakes have already been accelerated to a eutrophic stage. Where this is true, efforts of restoration may be considered as explained in the next section.

CONTROL AND RESTORATION

In the preceding sections, we have seen that there are many, many overlapping processes that change ground and surface waters adversely. Possibilities for limiting these processes from the start have been recognised, and restoration procedures have been identified. We will consider these possibilities and procedures next.

Each region, when confronted with contamination of soil, ground water, rivers, lakes, estuaries, and coastal waters, must employ whatever tactics appear to be technically, economically, and politically feasible.

These tactics may be directed at preventing future contamination, restoring already contaminated systems, restricting use of contaminated areas, and compensating those affected adversely. Let us first examine control measures and then consider restoration measures.

Control of Point-Source Contamination

Contamination of ground water takes place at countless dumps, landfills, disposal pits and ponds, and injection wells. Contamination of surface waters takes place wherever an outfall pipe discharges contaminants into rivers, lakes, and estuaries.

These sources are point sources that can be controlled. We will look at some of the control measures that can be taken in the siting of such facilities.

Sanitary landfills and various kinds of disposal pits are the most common sources of ground-water contamination. Siting of new landfills or pits is therefore a subject of intense interest in thousands of communities. The Babylon, Long Island, landfill is an example of a sanitary landfill; and the pits at Rocky Mountain Arsenal and the Love Canal disposal sites are examples of toxic-waste disposal sites.

An examination of sanitary-landfill siting and design will serve as a guide to the general principles to be considered for all of these facilities.

A *sanitary landfill* is a site at which domestic and municipal wastes are placed so that they do not create a public nuisance and do not create hazards to public health. In practice, wastes are trucked to the site, dumped onto the ground surface in a trench or against an embankment, and covered with some kind of earth materials.

As the landfill grows, it typically takes the form of a truncated pyramid rising above the surroundings. Within this pyramid are cells consisting of refuse, each more or less completely separated from other cells by the earth materials used as cover.

One of the best introductions to the technical details of landfill siting and hydrogeology is in the work of the Illinois Geological Survey. One of their reports is a major source for the following material.

The potential for contamination from landfills is determined by the kinds of wastes contained in them and by decomposition processes acting on the wastes. Decomposition is accomplished by microorganisms that act under aerobic conditions in the uppermost part of the landfill and under anaerobic conditions in the lower parts.

Carbon dioxide and methane are produced by this decomposition, and the methane migrating upwards is an important contaminant. A liquid, called *leachate,* forms from the refuse and contains high concentrations of solids and organic materials.

When precipitation infiltrates a landfill, the leachate is carried downwards to the water table. This is an intermittent process. If the base of the landfill intersects the water table, then the leachate can enter the ground water continuously. Therefore the goal in the design of landfills and disposal sites is to maintain release of the leachate or liquids at an acceptable level and to place the site so that escaping contaminants will be attenuated or will follow paths that avoid contaminating the surrounding areas. Hughes, Landon, and Farvolden have summarised the tactics needed for achieving these goals:

1. Eliminate the production of leachate.
2. Allow the leachate to migrate within an acceptable area.
3. Provide for recovery of the leachate at a distance from the landfill.
4. Retain the leachate so that it can be recovered. Let us examine how each of these tactics may be implemented.

Elimination of Leachate Production

If the infiltration of precipitation into the landfill can be controlled, leachate production may be partially controlled as well. The most economical way to achieve this is to cover the upper surface with the least permeable materials available.

Such materials may be glacial till, clay-rich materials, or even man-made cover such as asphalt. Such measures do not totally prevent the formation of leachate, since ground water may enter the landfill even if no water infiltrates from above.

Therefore it is desirable to situate landfills at sites above the water table. Such sites are common in arid regions, but they are scarce in humid regions.

Migration of Leachate Under Acceptable Conditions

Since most landfills produce and release leachate, it is important to situate them so that the leachate will move slowly and become decontaminated as it moves. The Illinois studies have shown that clays, glacial tills, and unfractured shales make excellent landfill substrates.

Ground water moves very slowly through these materials, and total dissolved solids are often reduced by a factor of 10 to 100 as the leachate travels a distance of a few meters.

Where materials having these properties are not to be found, then other approaches must be considered. At sites like that at the Babylon landfill, where the materials are highly permeable sands, one option is to place the site so that the flow paths terminate at a body of surface water where dilution may result.

At Babylon, the flow paths may pass beneath Santapogue Creek and terminate at Great South Bay. Another option is to place the site so that flow paths guide the leachate into a deep aquifer containing water that is known to be of unacceptable quality to begin with.

Migration and Recovery

Where local flow systems are well understood, it may be feasible to locate a landfill so that the leachate will be carried along flow paths converging at a discharge point where the leachate can be captured as needed.

For example, a landfill placed on a hill above flow lines terminating in a local swamp or pond might contaminate that swamp or pond without affecting the larger regional ground-water system.

If monitoring along the flow path showed that contamination levels were reaching unacceptable levels, leachate collection wells could capture the leachate for treatment.

Retention and Recovery

A landfill or pit can be designed so that a leachate recovery system is in place from the beginning of use. Such a system might consist of buried perforated pipes, called *tiles,* laid horizontally, or wells placed in or around the site.

These perforated pipes and wells could be pumped to withdraw the leachate. As they are pumped they would also alter the local flow system, preventing the leachate from entering the larger ground-water system. This tactic is illustrated in a cross section that shows the flow system under natural conditions and the flow system that would result from pumping from a buried tile system.

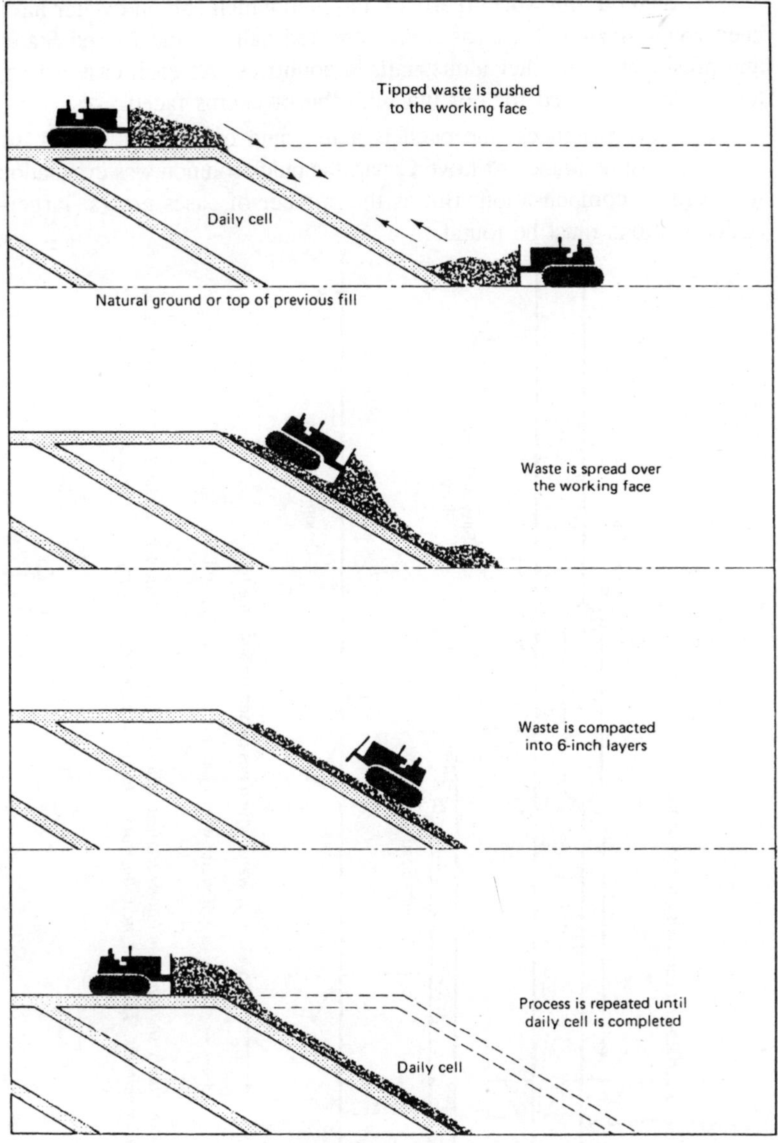

Figure 6.5 : Schematic diagram of a sanitary landfill showing cells surounded by cover material.

Restoration of Contaminated Systems

Throughout the world there are cases in which soil and water have been contaminated. New cases are reported daily in the United States and presumably in other industrialised countries. As each case arises, there may be a need for dealing with the problems faced.

In many instances, the need is acute, and quick solutions to the problem must be found. At Love Canal, the quick solution was evacuation followed by compensation. But as the number of cases grows, larger-scale solutions must be found.

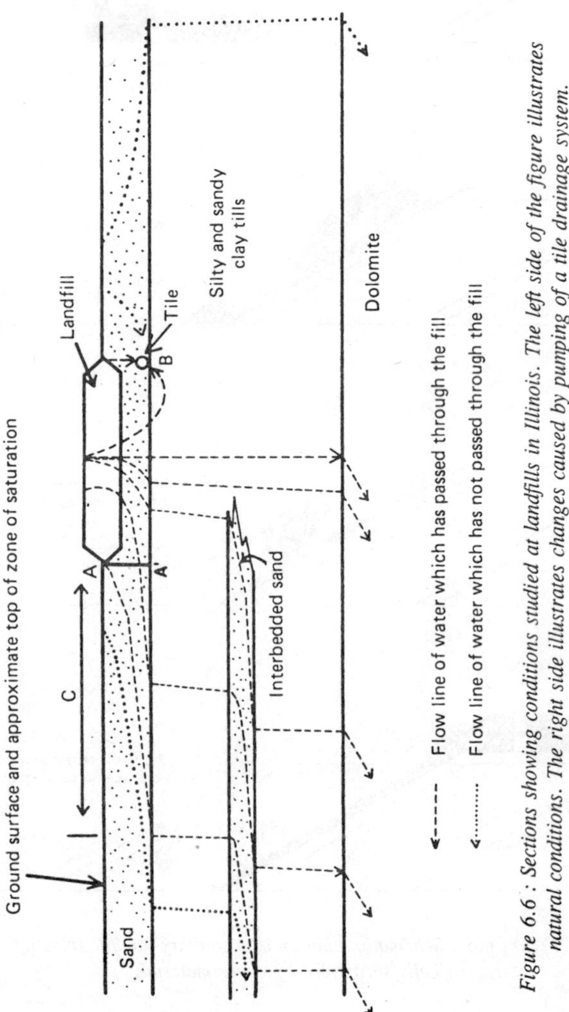

Figure 6.6 : Sections showing conditions studied at landfills in Illinois. The left side of the figure illustrates natural conditions. The right side illustrates changes caused by pumping of a tile drainage system.

Some of these will require the development of programs to restore and rehabilitate soils, aquifers, and surface waters. Others call for the development of programs of compensation.

In most instances, these programs demand a legislative and financial base that is created regionally or nationally. Here we will consider different kinds of restoration and rehabilitation. Policies that make these programs possible will be considered in the concluding sections of the chapter.

Examples of Ground-water Restoration

Once a contaminant has spread through a large volume of an aquifer, that contaminant can be removed or diluted only slowly, given the slow rate of ground-water flow in most systems. Yet some efforts have been undertaken and others surely will be.

At Love Canal, New York, the New York State Department of Environmental Conservation employed one of the tactics described in the preceding sections. Trenches were dug around the disposal site, and drainage tiles were placed in these.

Contaminated ground water moving into these tiles was pumped from them, and this water was then fed to a water-purification system in which activated charcoal was used to remove most of the contaminants from the water.

The purified water was then fed to the Niagara Falls municipal treatment system. This system is illustrated in Fig. elsewhere in this chapter and is described in detail in the U.S. Environmental Protection Agency report.

The Environmental Protection Agency has subsequently monitored wells around Love Canal for the various organic chemicals that had been found there earlier. The EPA found that the drainage-tile water-purification systems had apparently contained the spread of contamination.

The initial cost of this system was large ($8 million), and the costs of continuing to operate it are substantial. Thus, although this tactic can work, it is expensive. Furthermore, the toxic chemicals captured on charcoal must also be disposed of safely to avoid simply transferring the problem to a new site.

The system at Love Canal was installed after a large-scale evacuation. A similar approach is being followed at another Niagara Falls disposal site, but under somewhat different conditions. The site is the Hyde Park site north of Love Canal, a site also operated by the Hooker Chemical Corporation. After a court case brought by the U.S. Government against

Hooker, an agreement was made that stipulates in detail the nature of the monitoring system that is to be placed around the site, the leachate collection system that is to be installed, and the conditions under which this system is to be operated.

At Rocky Mountain Arsenal, where the source of contamination consisted of liquid wastes in open pits, a different combination of tactics has been employed. Here, the first step was the construction of new pits having impermeable liners.

The waste was pumped from the older pits into these new pits. This stopped the further infiltration of contaminated water. Then the older pits were filled with fresh water and kept filled to provide the hydraulic head needed to drive fresh water through the surface aquifer. Eventually this fresh water should dilute and displace contaminants from the aquifer. In addition, buried walls were constructed to impede the flow of ground water. Wells were placed on the upstream side of these walls to capture contaminated water.

This water was treated, and purified water was then recharged through wells on the other side of the walls. The predicted effects of this approach have been described by Konikow. The observed effects have been described by Konikow and Thompson.

These examples show that, to a degree, ground water can be restored. This requires considerable information about ground-water flow and substantial financial support. Restoring contaminated soil and ground water is much more costly than preventing contamination.

Examples of Lake and Stream Restoration

Substantial efforts are underway to restore the water quality of lakes and streams in the United States. For many rivers the first goal has been to ensure that dissolved oxygen levels are raised to former levels and are maintained there.

A second goal has been to deal with contamination by substances such as polychlorinated biphenyls (PCBs) or dioxin. For many lakes the initial goal has been to slow the process of induced eutrophication by controlling algal productivity. An example of each of these efforts is presented here.

The Willamette River in Oregon is one of the most carefully studied rivers in the United States. Each of a continuing series of U.S. Geological Survey Circulars 715A-K concerns various aspects of river quality assessment of the Willamette River Basin. As explained by Rickert and collaborators, "The maintenance of high dissolved oxygen concentration has been the critical problem in the Willamette River".

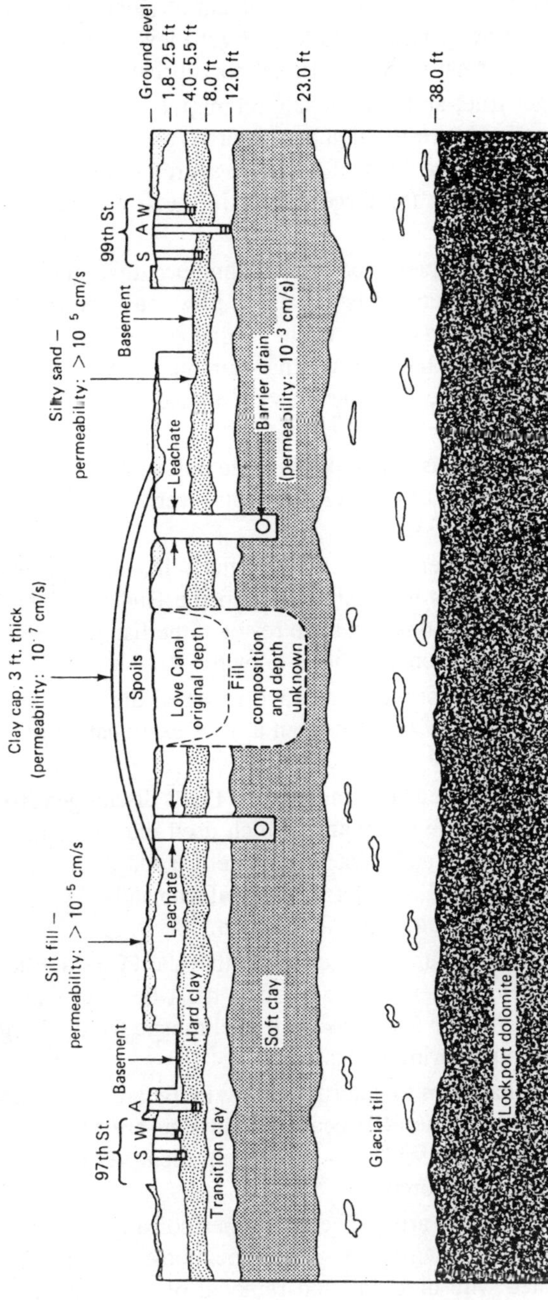

Figure 6.7 : Remedial drainage system at Love Canal. Buried drainage tiles capture leachate from teh disposal site.

Until rather recently the dissolved oxygen content of river waters was very low during low-flow periods, and low or zero levels of dissolved oxygen hindered the normal migration of salmon upriver.

This is no longer true, and normal salmon migration patterns have been partially restored. This restoration has occurred because a program of secondary wastewater treatment of all point-source discharges was instituted. The Willamette is the largest river in the United States to be so fully controlled.

The USGS has developed modeling techniques that allow the prediction of dissolved oxygen levels at various flow rates and to study the effects of the augmentation of flow or the further control of industrial discharges. This modeling has shown that summer flows must be augmented in order to assure meeting present standards for dissolved oxygen.

Any further improvements will depend on the reduction of industrial waste inputs of ammonia in the stretch between river mile 50 (above the mouth) and river mile 134.

The improvement in water quality seen for the Willamette has been seen in many other rivers as well, among them the Hudson River in New York. Improvements of this sort have restored the fish populations for which these rivers were once so well known.

In the process, these improvements have revealed the importance of other contaminants, as may be seen from a brief examination of the Hudson River.

The Hudson river has been contaminated by PCBs discharged from plants of the General Electric Company, which used the PCBs in the manufacture of transformers and capacitors. The use and discharge of PCBs are now forbidden by U.S. federal law. Unfortunately, PCBs are long-lasting and continue to enter the food chain.

Studies of water and sediment have shown that the PCBs are held by the sediment particles. Therefore two issues must be dealt with in efforts at restoration. The first issue has been to eliminate the discharge of PCBs from the manufacturing plants.

The second issue has been to determine how PCBs are moved and buried in river sediments. If the PCB-bearing sediments are covered by uncontaminated sediments and will remain covered, then they may pose little threat and might be ignored.

But if flood discharges are expected to periodically erode the sediments and move them downstream to new locations, then the costs and benefits associated with dredging and removal of sediments from the river should be considered. The first prerequisite is to obtain good information about sediment concentrations and sediment movement patterns. Preliminary findings of research on the Mohawk and Hudson have been published, but final resolution lies in the future.

The possibility for restoration of lakes and the complexity of restoration are particularly well illustrated by efforts to bring about improvements in water quality of Irondequoit Bay in Rochester, New York. This bay was undergoing cultural eutrophication, probably increased sediment loading, and remarkable loading by sodium chloride during the period from 1950 through the 1960s.

As a result of these changes, it was a target for scientific research on eutrophication, salinity change, and restoration. The restoration measures have taken the following forms. In 1973, New York state banned the use of phosphate in household detergents.

Then a large-scale system for phasing out all sewage treatment plants discharging effluent into this bay was planned and constructed. Finally, the use of de-icing salts was reduced within the watershed of the bay.

These control measures have brought about decreases in the loading of nutrients and in the chloride-ion concentration in bay waters. In spite of these reductions, additional sources of contamination have been identified. One source is the lake-bottom sediments themselves.

Sediments entering in past decades have trapped nutrients on the bottom of the lake. Will the lake sediments release nutrients and other contaminants to lake waters as the input of contaminants from external sources is decreased? The answer is not yet known; research leading up to this is reviewed in Bannister and Bubeck.

According to studies done by the U.S. Geological Survey under the National Urban Runoff Program (NURP), water quality in the bay has not improved as much as expected. This program is investigating the contamination contributed by surface runoff and ground water within the watershed.

RADIOACTIVE WASTES-PLANNING FOR THE FUTURE

Prediction of future risk has not been an important element in the selection of sites for the disposal of most nonradioactive wastes. Thus there are few examples of scientific prediction in the records of site selection and design for the hundreds of sites identified by the EPA as sources of concern.

In contrast with this record, planning for the disposal of the radioactive wastes created by commercial nuclear-fission reactors is characterised by very large-scale, long-term efforts to employ scientific analysis and prediction.

This is true in all countries in North America and Europe that have nuclear-reactor programs, and all of these countries are investing substantial scientific resources in the design of their programs for radioactive-waste management. For this reason, the scientific basis for radioactive-waste management is worth special examination.

What is radioactive waste? Why is it accorded such unusual attention? To answer the first question requires a familiarity with the fission-reactor approach to the generation of electricity. To answer the second question requires a familiarity with the growth of knowledge about the human-health effects of ionizing radiation.

Each of these topics is treated briefly here before examining the actual process of scientific analysis of the problem of nuclear-waste disposal.

The Generation of Radioactive Waste

Nuclear-fission reactors rely on the controlled fission of uranium 235. Different nuclear reactor designs are employed, but all employ reactor *fuel elements* made up of mixtures of uranium 235 and uranium 238. The fuel elements consist of uranium mixtures fabricated into cylindrical pellets or into cubes.

These pellets or cubes are loaded into the reactor and then become the sites of the controlled chain reactions on which nuclear-fission power depends. The operation and design of these reactors will not be discussed here. An excellent source of information on this subject is Nero.

The basic reaction on which all fission reactors depend may be shown in a simplified diagram. This diagram shows a neutron hitting a single uranium 235 nuclide and including fission in this nuclide. This is called *fission* because each individual event produces a pair of fission products, nuclides with atomic numbers that are between 40 and 65. This fission also produces neutrons and releases energy.

The released energy is used to heat a fluid that may be water, heavy water, helium, or carbon dioxide. The fluid transfers heat from the reactor core to the turbines, where electricity is generated. Some of the released neutrons interact with the fluid or the reactor materials; others induce fission events in other uranium 235 nuclides. Still others interact with uranium 238 to produce plutonium 239.

The fission process produces a vast array of fission products and other nuclides. These other nuclides include tritium, carbon 14, and the actinides, which are those elements having atomic numbers from 90 through 103. All of these newly produced nuclides are radioactive.

Thus a nuclear-fission reactor is a device that *produces* radoactive nuclides as well as energy. In this chapter we will examine the reactor output of radioactive nuclides, which eventually become part of the radioactive wastes that must be disposed of, and we will trace the possible methods for disposal of these wastes.

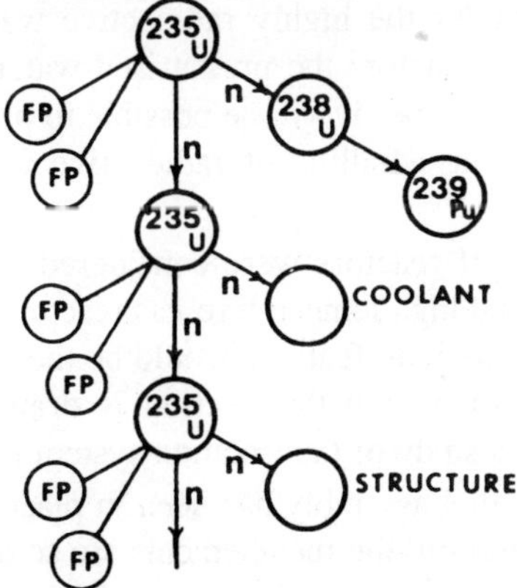

Figure 6.8 : Schematic diagram of fission reaction of ^{235}U producing fission products (FP), neutrons (n), and energy. Diagram also shows possible interactions between the released neutrons and various materials.

Characteristics of Nuclear-Reactor Waste

The exact character of nuclear-reactor wastes is determined by the exact nature of the nuclear reactor and by the nuclear-fuel cycle that is employed. Alternative nuclear-fuel cycles are illustrated in Figure elsewhere in this chapter. In the *once-through cycle,* ore is mined, milled, and sent to an enrichment plant.

The enriched ore is then used in the fabrication of the fuel elements. These fuel elements are placed in fuel assemblies, which are then placed in the reactor. When the fuel assemblies are removed from the reactor, they are sent on for disposal as intact fuel assemblies.

In the oncethrough cycle, any unused uranium 235 present in the fuel elements at the time of removal is simply lost to further use. Since this unused uranium 235 has potential value as a fuel, new systems have been devised to extract it from the fuel assemblies. These systems are based on a reprocessing fuel cycle.

In the *reprocessing cycle,* all the steps are the same until after the assemblies have been removed from the reactor. The assemblies are then sent to a reprocessing plant where the unburned uranium 235 and the plutonium 239 that has been produced from uranium 238 during the 880 days of operation are removed.

This cycle has been employed in the United States in the past, is being employed in other countries, and may be employed in the United States in the future. Therefore it is important to examine it here.

Radioactive Waste from a Once-through Cycle

Here we consider the highly radioactive waste produced by one type of light-water reactor, the pressurised-water reactor (PWR). By considering only one type, it will be possible to present the information needed to discuss the handling of radioactive waste without creating undue complexity.

If other types of reactors were considered, such as boiling-water reactors (BWRs) and high-temperature-gas reactors (HTGRs), the details would differ, but the main features would be the same as for the PWR. Most of the information in this section is drawn from the National Research Council's study of the isolation system for radioactive wastes.

After a PWR fuel assembly has been in place for 880 days of full-power reactor operation, the fuel elements are so depleted in fissionable uranium and so contaminated with reaction products that the assembly must be removed from the reactor.

At the time of removal, a single fuel assembly contains fission products, activation products, and actinides, which make it intensely radioactive. This radioactivity is a source of health hazards and of heat. Let us consider the heat production first.

A just-removed fuel assembly generates heat at a rate of 10,000 watts. This heat is generated largely by the decay of fission products that have relatively short half-lives. After removal from the reactor, heat production declines.

At the end of 100 years, the heat output from a PWR assembly has declined from 10,000 watts to 100 watts. Thus heat production is obviously a major factor to be dealt with in handling and storing fuel assemblies in the first decades after removal.

After 300 years, the heat produced comes largely from decay of the actinides. This suggests that the nature of the radiation hazard changes with time. Let us look at the inventory of radioactive nuclides in a PWR assembly to see what contributes to this hazard.

Table 6.2 : Partial Inventory of Radioactive Nuclides fromOne PWR Fuel Assembly at 10, 100 and 1000 Years After Removal.

	Time after discharge					
	10 yrs		100 yrs		1000 yrs	
Nuclide	grams	MBq	grams	MBq	grams	MBq
^{14}C	$1.6 \times 10^{-'}$	2.5×10^{4}	$1.6 \times 10^{-'}$	2.5×10^{4}	$1.4 \times 10^{-'}$	2.3×10^{4}
^{79}Se	2.7	7.0×10^{3}	$2.77.0 \times 10^{3}$	2.7	6.9×10^{3}	
^{90}Sr	2.0×10^{2}	1.0×10^{9}	2.3×10^{1}	1.1×10^{8}	-0	-0
^{99}Tc	3.6×10^{2}	2.3×10^{5}	3.6×10^{2}	2.3×10^{5}	3.6×10^{2}	2.3×10^{5}
^{126}Sn	1.3×10^{1}	1.3×10^{4}	$1.3 \times 10'$	1.3×10^{4}	$1.3 \times 10'$	1.3×10^{4}
1291	$8.3 \times 10'$	5.3×10^{2}	$8.3 \times 10'$	5.3×10^{2}	8.3×10^{1}	5.3×10^{2}
^{135}Cs	1.4×10^{2}	6.3×10^{3}	1.4×10^{2}	6.3×10^{3}	1.4×10^{2}	6.3×10^{3}
^{137}Cs	4.4×10^{2}	1.4×10^{9}	5.5×10^{1}	1.8×10^{8}	-0	—0
^{226}Ra	1.6×10^{-7}	5.4×10^{-3}	1.2×10^{-5}	4.2×10^{-1}	1.4×10^{-3}	$4.9 \times 10'$
234U	$8.8 \times 10'$	2.0×10^{4}	1.2×10^{2}	2.7×10^{4}	1.5×10^{2}	3.3×10^{4}
238U	4.4×10^{5}	5.4×10^{3}	4.4×10^{5}	5.4×10^{3}	4.4×10^{5}	5.4×10^{3}
^{237}Np	2.1×10^{2}	5.4×10^{3}	2.7×10^{2}	7.2×10^{3}	6.6×10^{2}	1.7×10^{4}
^{238}Pu	6.0×10^{1}	3.7×10^{7}	3.0×10^{1}	1.8×10^{7}	2.6×10^{2}	1.8×10^{4}

(Table Contd.)

(Table Contd.)

Nuclide	*Time after discharge*					
	10 yrs		100 yrs		1000 yrs	
	grams	MBq	grams	MBq	grams	MBq
^{14}C						
^{239}Pu	2.3×10^3	5.3×10^6	2.3×10^3	5.3×10^6	2.3×10^3	5.1×10^6
^{240}Pu	1.1×10^3	8.7×10^6	1.1×10^3	8.7×10^6	1.0×10^3	7.9×10^6
^{241}Pu	3.5×10^2	1.3×10^9	4.6	1.8×10^7	8.6×10^{-5}	3.6×10^2
^{242}Pu	2.1×10^2	3.1×10^4	2.1×10^2	3.1×10^4	2.1×10^2	3.1×10^4
^{241}Am	2.3×10^2	2.9×10^7	5.0×10^2	6.4×10^7	1.2×10^2	1.5×10^7
^{243}Am	$4.0 \times 10'$	2.8×10^5	3.9×10^1	2.8×10^5	$3.6 \times 10'$	2.6×10^5
^{245}Cm	5.6×10^{-2}	3.5×10^2	5.5×10^{-2}	3.5×10^2	5.1×10^{-2}	3.3×10^2
^{246}Cm	4.3×10^{-2}	4.9×10^2	4.3×10^{-2}	4.9×10^2	3.7×10^{-2}	4.2×10^2

At the time of discharge, fission products are major contributors to the output of ionizing radiation from the fuel assemblies. This may be seen from a table listing a partial inventory for a PWR at 10, 100, and 1000 years. This table shows that fission products such as strontium 90 and cesium 137 are major sources of radiation during the first 100 years, and, in fact, during the first 600 years.

By the time 1000 years have elapsed, however, these fission products are negligible sources of radiation and heat. As the radioactivity from most fission products declines, the radioactivity from the actinides increases; and after a few hundred years, most of the radiation output comes from americium, neptunium, plutonium, and two important long-lived fission products, technetium 99 and iodine 129.

The radioactivity produced by each of these elements may also be seen in Table elsewhere in this chapter. These elements present a source of radiation hazard for very long periods-of time.

One way to appreciate the degree of health hazard caused by these various elements is to examine a diagram that shows the volume of water that would be required at any given time if one were to mix a specific nuclide with sufficient water to create a solution that could be ingested safely.

Such a diagram shows, for example, that the amount of strontium 90 present in a reactor assembly containing one megagram (10^6 grams) of heavy metals (uranium, plutonium, etc.) would have to be diluted in more than 10^{10} cubic meters of water to meet the health standard. The fission-product curves in the diagram show that the volume of water required to dilute most of the fission products to safe levels declines to zero after periods of 10 to 1000 years have elapsed.

In other words, most fission products are not a source of hazard after periods ranging from 10 to 1000 years.

Radioactive Waste from Reprocessing

The fuel assemblies described above are the form that reactor waste takes if a *once-through cycle* is employed. These assemblies may be reprocessed by dissolving them in nitric acid and extracting uranium 235 and plutonium 239 for future use. This extraction leaves most of the fission products and the actinides in the nitric acid solution.

The extraction process also frees such gaseous or potentially gaseous elements as carbon, hydrogen, krypton, and iodine from the fuel elements. These must be captured by appropriate techniques, or they will escape to the atmosphere.

Reprocessing creates a collection of radioactive wastes including an acid solution, here referred to as high-level waste (HLW), and separated masses of iodine, hydrogen 3 (tritium), carbon, and krypton, as shown in Figure elsewhere in this chapter. All of these wastes must be dealt with, but since they are in different forms, special waste-handling techniques must be employed. Perhaps the most significant technique is that for handling the nitric acid solution.

To date, commercial wastes from reprocessing have been stored in steel tanks at West Valley, New York. These wastes and others like them must be solidified. This solidification process was begun at West Valley in 1984. The first step in the solidification process is the production of granular material through a process called *calcining*.

Then this granular, highly radioactive material is converted to a stable solid. Two forms may be noted here. The granular material might be subjected to fusion to produce meter-long cylinders of *borosilicate glass*. Alternatively, it might be heated with appropriate silicate materials to produce centimeter—long rods of SYNROC, a synthetic analogue of natural silicate rocks.

If the fusion technique is employed, the radionuclides are dispersed in solution in the borosilicate glass. If the SYNROC technique is employed, the radionuclides take up specific positions in the crystal structures of the synthetic minerals of which the SYNROC is made. In either case, a stable, reasonably strong, but highly radioactive cylinder is produced that is more easily managed than the nitric acid solution from which it was made.

Geologic Isolation Systems for Nuclear Waste

Basic Goals

Whatever the form taken by the radioactive wastes, they must be isolated from contact with the biosphere. All of the countries included in the Organisation of Economic Cooperation and Development (OECD) are at work on developing systems for the isolation of radioactive wastes in rock bodies beneath the earth's surface.

Progress in each of these countries is reported in an OECD study. Those who are studying these isolation systems must evaluate the problems to be solved in dealing with storage of fuel elements, borosilicate glass, and SYNROC, because it is possible that all three waste forms will be employed.

The waste will be placed below ground in a permanent repository developed in rock or sediment. This method of below-ground storage is called the *geologic approach* to radioactive-waste disposal. It has been adopted in the belief that it provides the best guarantee that radioactive wastes will be kept from contact with the biosphere over periods of thousands of years.

The decisions to be made concerning geologic disposal are land and water decisions of the first rank because they have the potential for affecting land and water use for thousands of years. Some of the criteria to be considered are:

1. The waste package should remain stable for at least 1000 years.
2. The waste package should be isolated from ground water as long as possible.
3. The waste package should be designed so that there will be the least possible exchange with ground water, if contact is made with ground water.
4. The ground-water path between the waste package and the biosphere should be such that radionuclides released from the waste package will not reach the biosphere for at least 1000 years after release.

Scientists and engineers are confronted with unprecedented prediction problems. They are being asked, for example, to provide scientifically based predictions concerning the path that released radionuclides will follow and the geochemical behaviour that these nuclides will display while moving along that path.

All such predictions will be "best-possible" predictions rather than certain predictions, so all efforts to design isolation systems include plans to provide *multiple barriers* to the ultimate dispersal of radionuclides. We will illustrate this by an example.

We will examine the actual state of investigations into geologic disposal systems. This will illustrate both the scientific prediction problems and the issues that must be faced.

Here we will look at progress in two countries, the United States and Sweden. Both countries use the same types of reactors, and both have national laws specifying that safe storage systems must be developed to hold reactor wastes.

The Swedish laws were passed in 1977; the U.S. laws were passed in 1982 (see below). We will consider Sweden first because it appears to be closer to actually instituting a long-term isolation system.

CHANGING ATTITUDES TOWARD CONTAMINATIONAL LAWS AND PRACTICE

During the years immediately following the Second World War, many bodies of surface water were allowed to become very seriously contaminated. By the late 1960s and early 1970s, changing attitudes began to be reflected in legislation to deal with this contamination.

The initial concerns were with surface water, but more recently concerns have been shown for ground-water and soil contamination. In what follows we draw upon the survey of water-quality laws provided by the Council of Environmental Quality in its annual reports to show the changing concerns with the contamination of soil and water.

Efforts to improve the quality of the surface waters of the United States began with passage of the Federal Water Pollution Control Act Amendments (FWPCAA) of 1972. This act specified that all municipal sewage treatment plants were to provide secondary treatment of sewage by 1977 and were to employ the best available technology (BAT) by 1983.

This law directed the Administrator in the Environmental Protection Agency to prepare or develop comprehensive programs for "preventing, reducing, or eliminating the pollution of the navigable waters and ground waters (33 USC sec. 1252a)" The Council on Environmental Quality observed that most cities had not met this goal by 1981 (Council on Environmental Quality, 1981). Thus the achievements reported here for the Willamette River illustrate what was intended by the law.

The FWPCAA was amended in 1977 by passage of the Clean Water Act, which was introduced to demonstrate the growing recognition that toxic contaminants from industrial sources were not being adequately controlled by the 1972 law. The Clean Water Act introduced a National Pollutant Discharge Elimination System requiring that each contributor to point-source discharge obtain a permit from the EPA and by so doing inform the EPA of the magnitude of character of discharge from each point source in the country.

Then it was recognised that even the complete control of point-source discharge from sewage treatment plants and industrial users might not be sufficient to achieve the desired end. This recognition was based on growing understanding of the importance of nonpoint-source contamination, which was not at all regulated and which was perhaps not easily controllable in any case. The EPA has no explicit authority to regulate nonpoint-source discharge, so it has chosen to operate under

Section 208 of the Clean Water Act to provide for research on the contributions of urban storm runoff and agricultural runoff and to provide aid to municipalities for establishing plans for controlling such runoff. In addition, the U.S. Department of Agriculture is operating model programs to show what might be accomplished by controlling agricultural runoff.

Although there is a long record of legislative efforts to deal with surface-water contamination, comparable efforts to deal with ground-water contamination have only begun. The EPA chose to concentrate on the then more immediate and readily demonstrable problems with surface water.

The problems of ground-water contamination were recognised with the Safe Water Drinking Act of 1974. This law requires that the EPA set drinking-water standards. The only actual provisions for exercising control over ground water are contained in the Underground Injection Control Program and in what is called the Gonzalez Amendment.

Research efforts have begun to deal with the Underground Injection Control Program and the Gonzalez Amendment. These are directed toward identifying major aquifers and developing hydrologic data and models for each. Given these findings, the EPA will be able to determine where underground injection of wastes must be sharply limited.

It will also be able to determine if a particular aquifer is what the Gonzalez Amendment calls *a sole-source aquifer.* Such an aquifer is one on which existing water users are totally dependent. The EPA has limited authority to prevent actions that might contaminate such an aquifer.

As contamination from hazardous disposal sites has become better understood and more widely recognised, legislation has been introduced to deal with this also. This legislation includes the Resource Conservation and Recovery Act of 1976 and the Toxic Substances Control Act of 1976, commonly referred to, respectively, as RCRA and TSCA. Both of these acts address the regulation of point-source discharge of toxic substances. They are not concerned with contamination that has been identified.

The one other major federal law that deals with ground-water contamination is the Surface Mining Control and Reclamation Act of 1977. This act recognizes that surface mining can have disruptive effects, both chemical and physical, on ground-water systems, and it spells out the efforts to be made to mitigate these effects.

7

VOLCANISM

Worldwide volcanic activity in the period 1973-1984 has provided examples of most of the kinds of eruptive activity that must be faced in dealing with volcanic hazards. In 1973, a newly created volcano on the island of Heimaey, Iceland, buried a town under a cover of black volcanic tephra. Mt. Etna in Italy became active in 1979.

In April 1982, El Chichon volcano in southern Mexico erupted, causing great loss of life and giving climatologists and meterologists reason to believe that the eruption might affect the earth's weather and climate. Kilauea volcano on the island of Hawaii began to erupt in January 1983 and released lava flows that threatened residential areas.

In 1984, Mauna Loa volcano, which is adjacent to Kilauea, also began to erupt. Within the conterminous United States, Mount Baker in northern Washington State became a source of concern when melting of glacier ice on the volcano was observed in 1975. Then, in 1980, Mount St. Helens in southwestern Washington State began an extended period of eruptive activity that caused great destruction and substantial loss of life.

At about the same time, Long Valley caldera and the Mono-Inyo craters area in California became targets of growing concern as potential eruptive centers. They were being watched closely through 1984. Experience with these and other eruptions showed that scientific information can provide a basis for warning and evacuation that can save lives.

It also led to some advances in our ability to assess volcanic hazards and to predict future eruptions. These eruptions were also of interest to those who have examined the historic relationships between humans and volcanic activity at Pompeii in Italy, Santorini in the eastern Mediterranean, and Krakatau and Tambura in the East Indies. Here we will examine volcanic hazards from a late 20th-century viewpoint by looking at research on the prediction of volcanic eruptions and the effects of eruptions.

We will also examine adjustments to volcanic hazard. First we consider the processes leading to eruption and then the eruptive events and products. We will draw on studies of three of the volcanoes cited above to illustrate the general principles and to provide background for a discussion of volcano-hazard assessment. These studies will show the adaptations that have been employed and that will be employed in the future.

THE ERUPTION PROCESS AND ITS PRODUCTS

All volcanic eruptions are rooted in a two-step process that precedes the eruptions. The first step is the melting of rock masses tens of kilometers below the earth's surface to produce *magma,* which is molten rock containing dissolved water and other gases. The second step is the movement of this magma toward the earth's surface.

The first step determines the initial temperature and composition of the magma; the second step controls the nature and timing of eruption. Let us consider the material that is produced. Then we will examine the main stages in the volcanic process. Some general sources that may be consulted for amplification include: MacDonald (1972), Sheets and Grayson (1979), and Tazieff and Sabroux (1983).

We will use names designating four types of magma, which can be distinguished on the basis of their content (weight-percent) of silica (SiO_2): (1) basaltless than 50 percent silica; (2) andesite-52 to 55, percent silica; (3) dacite-60 to 65 percent silica; and (4) rhyolite-more than 70 percent silica.

These magmas and the rocks formed from them are not generally defined on this criterion alone, but this will suffice for the purposes of this chapter. Magmas are very hot liquids; temperatures range from as low as 600°C to 1250°C. These magmas differ greatly in the ease with which they will flow and in the ease with which the dissolved gases can escape from them.

In general, basalt magma, which is relatively low in silica, flows most easily and is therefore said to have the lowest *viscosity*. Rhyolite magma, which is silica-rich, flows less easily and has the highest viscosity. These viscosity differences are important in controlling the physical form that eruptions take. For example, a low-viscosity magma may be erupted in lava fountains and lava flows, whereas a high-viscosity magma rarely displays these forms of eruption.

The gases dissolved in magma are especially important in determining the physical behaviour of the magma as it nears the surface and is erupted. It appears that andesitic, dacitic, and rhyolitic magmas generally contain much higher concentrations of dissolved gases than basaltic magma.

In general, water constitutes more than half of the total dissolved gases. The composition of the magma erupted at a particular volcano or fissure is determined by the nature of the rock being melted, by the depth at which melting takes place, and by processes that take place as the magma interacts with other rocks and fluids.

Studies of the rocks erupted at many different volcanoes show that magma has been generated for thousands or millions of years at sites deep beneath these volcanoes. In oceanic settings like Hawaii or Iceland, the magma, which is typically basalt, is generated within the upper mantle of the earth. In subductionzone settings like the Aleutians or the Cascades of the northwestern United States, andesites, dacites, and rhyolites may be generated by the partial melting of crustal rocks overlying the mantle.

In continental settings away from subduction zones, such as those represented by Valles Caldera in New Mexico, or Long Valley caldera in California, magmas commonly are silica-rich dacite or rhyolite. In all these settings, melting is the initial step. It is followed by a second multistage step in which the magma migrates upwards, moving through conduits that may be pipelike or may be planar fractures.

During this second step, the magma may move directly to the surface; more commonly it moves to intermediate levels, where it collects in subsurface reservoirs. The magma may move from such a reservoir to the surface, where it is erupted, or it may move to yet another reservoir even closer to the surface. Magma may solidify in a conduit or a reservoir, forming a body of *intrusive rock*.

A view of a simple eruptive system is illustrated here for the Hawaiian Islands. This shows that the initial melting process takes place far below the active volcanoes such as Kilauea and Mauna Loa. The eruptive system beneath a subducting plate is more complicated, as is suggested by Figure elsewhere in this chapter, which shows a series of feeder pipes above a subducting plate. It also shows that the magma

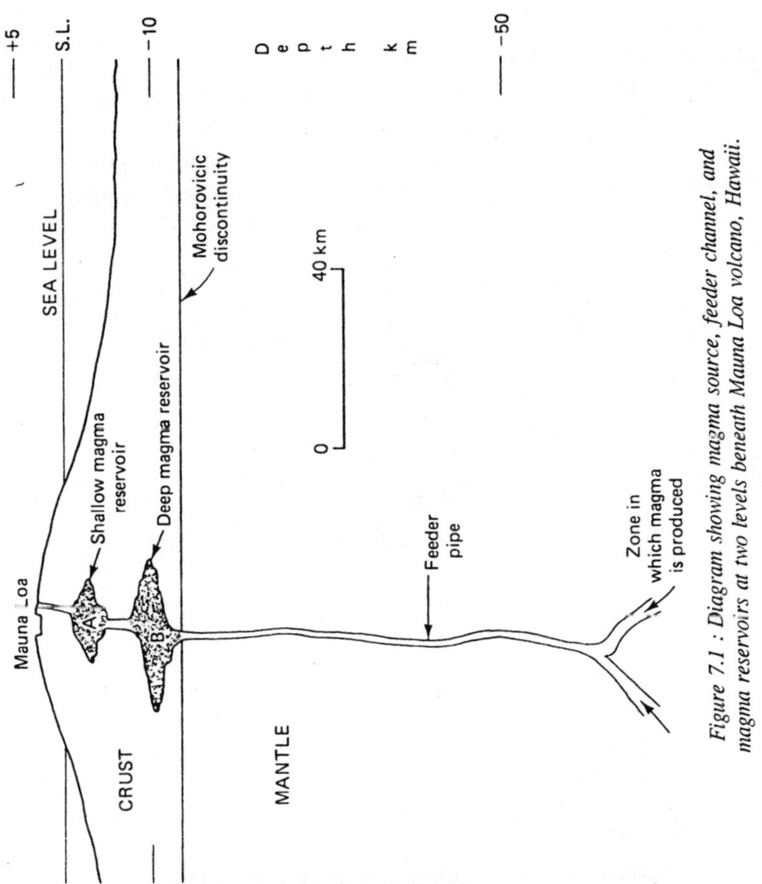

Figure 7.1 : Diagram showing magma source, feeder channel, and magma reservoirs at two levels beneath Mauna Loa volcano, Hawaii.

collects in chambers or reservoirs before being released. More complicated systems are illustrated throughout the chapter.

Throughout the time interval between initial melting and the actual eruption, magmas undergo changes in chemical composition. These processes of chemical and physical evolution produce magmas that differ in composition and physical behaviour from the original magma. This evolution commonly is reflected in the sequential eruption of different kinds of magma from a single volcano.

This sequence also leads to changes in the kind of eruptions observed over time. The role that volcanic activity plays in contributing to volcanic hazard depends on the rate at which the eruption takes place and on the scale of the eruption. Some eruptions take place slowly and affect limited areas.

These eruptions do not cause much loss of life, even if close to human settlements. Other eruptions occur with great rapidity and affect large areas within minutes. These can cause great loss of life, even in areas at some distance from the volcano. Let us examine the principal kinds of eruptions and look at their products. During an eruptive event, the material fed to the surface has a general chemical composition indicated by its name-basalt, andesite, and so on.

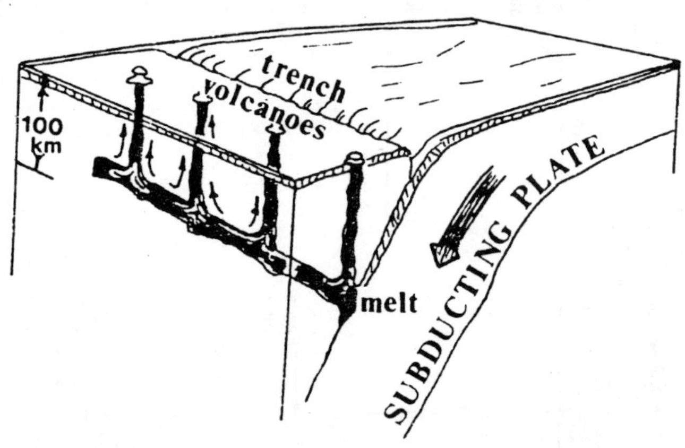

Figure 7.2 : Magmatic feeder pipes above a subducting plate.

The magma may also contain dissolved gases that separate out, at or near the surface of the earth, as a separate vapor phase. This vapor forms bubbles within the magma. Thus, at the moment of eruption, the material emitted is a mixture of molten magma, gases, and fragments of solid rock.

The character of the eruption depends on the relative abundances of these materials, on the total supply available, and on the geologic setting. Erupted material that consists largely of molten rock is called *lava.* Highly fluid lava may be erupted in fountains, such as those displayed at Kilauea on the island of Hawaii or in Icelandic volcanoes. These fountains are sites at which gases escape from the magma. The lava may also flow across the surface, forming *lava flows.*

Highly viscous lavas may form short and stubby flows that pile up on one another in the crater of the volcano, forming *domes.* The less viscous lavas form flows that may extend many kilometers from the source. Lava ordinarily moves about as fast as a person can walk. On steep slopes, however, lava can move much faster. When activity stops, each flow remains on the surface, forming a deposit that represents a net addition to the landscape.

Where an eruption is driven by the expansion of gases dissolved in the magma or by superheated ground water flash-boiling into steam, eruptions take on a different character. These eruptions are highly explosive. If the expanding water vapor comes from ground water in the pores of the rock of which the volcano is made, the eruption is called *a phreatic eruption.* During these explosive eruptions, hot fragmental material is ejected from the volcano.

The same material may be ejected during less explosive eruptions as well. Two general terms are used interchangeably for this fragmental material: *pyroclastic material* and *tephra.* Tephra will be the term generally used here. When fragmental material is ejected, it consists of blebs of still molten magma and crystalline rock fragments. The molten magma blebs quickly chill into glass.

The crystalline rock fragments consist of older rocks torn from the walls of feeder pipes and the upper part of the volcano. These fragments range in size from blocks much larger than a loaf of bread down to particles that are only fractions of a millimeter in diameter. Some of the larger pieces have streamlined forms and are sometimes called *bombs.* The smaller particles are called *lapillae, ash,* and *dust.*

Much of the finer material is produced when the expanding gas bubbles explode the hot glassy material within which they are at first contained. We see evidence of this process in the rocks called *pumice* and *vesicular basalt,* which contain closely spaced spherical and tubular cavities surrounded by glassy rock. In these rocks the bubbles were growing but not so rapidly or forcibly that they could shatter the surrounding rock.

Where this kind of bubble growth occurs more forcibly, the enclosing rock is shattered to form pyroclastic material. Tephra may be erupted vertically, rising in great clouds above active volcanoes. This airborne tephra may then drift downwind and settle out in deposits that are thickest closest to the source.

However, during some types of eruptions, fragmental material follows a path down the slopes of a volcano, remaining close to the ground as it moves. Many different names are used to describe these eruptions and the material driven out and deposited by them. Some of the terms used to describe the eruptive activity include *lateral blast, directed blast, pyroclastic flows and surges,* and *nuees ardentes.*

In all instances the moving material is hot, contains entrained gases, and moves very rapidly. This kind of activity is the source of the worst volcanic hazard. A geologic record of volcanic activity in any region exists in the form of lava flows, ash falls, cinder cones, and tephra deposits. The landscape may also provide a record of the centers and fissures from which these eruptions have been generated.

On mid-ocean islands like Hawaii, these centers are represented by volcanic craters and by fissures that open on the flanks of the a volcanoes. In other settings, such as at Mono Lake in California, the relatively flat terrain may be dotted with small cinder cones. Some volcanic areas are also marked by large circular depressions called *calderas*.

Such structures mark the sites at which a very large volume of lava has been erupted in a very short period of time. This sudden expulsion of so much magma leads to the collapse of a circular rock mass to form the depression we call the *caldera.* In regions such as the active subduction zones around the rim of the Pacific, the volcanic record is extremely complicated. It is dominated by great stratovolcanoes built up of interleaved lava flows and tephra.

The rocks in these volcanoes are dominantly andesites and dacites, but basalts and rhyolites may also be present. Within regions of older subduction-zone activity, there are also thick and areally extensive deposits of dacite or rhyolite that formed entirely from fragmental material discharged in fast-moving ash flows.

Some of the largest of these are associated with giant calderas, and they represent the material erupted from the reservoir beneath the site of the caldera. These are the basic forms that eruptive activity can take. These forms will become more familiar through a consideration of the case histories.

These case histories are presented for a very specific purpose: To show clearly the contribution of the different forms of eruption to volcanic hazard.

VOLCANIOC HAZARD EVALUATION

The evaluation of volcanic hazards had been given substantial attention in the United States, even before the eruption of Mount St. Helens. Since Mount St. Helens erupted, even more attention has been given to volcanic hazards, as may be seen from the review of the U.S. Geological Survey's research on volcanic hazards. In addition, natural and social scientists have assembled a record of the effects of volcanic activity on human ecology.

This book is an excellent guide to the literature and is an important source for what follows.The three cases we have just examined show that the dimensions of volcanic hazard depend not only on the nature of the potential volcanic events but also on the economic resources of the country or region where the volcanic hazard exists. This is shown vividly by comparing the situation at Mount St. Helens with that of El Chichon.

The events are similar, but the settlement patterns and economic dependence of the people on land use close to the volcano make the events more devastating in human terms at El Chichon. This is true for all hazards, but it is worth reiterating here that the physical examination of potential events dealt with in this section is only the first step in hazard evaluation.

A consideration of the second step, a study of adaptations, is treated in the concluding section of this chapter.

Identification of Centers of Potential Volcanic Activity

We have already seen a capsule view of the forms of the prediction of volcanic events and their effects. Here we examine the subject on a larger scale and look at two forms of areal prediction. One is concerned with identifying future centers of eruptive activity; the other is concerned with identifying the areas that will be affected by that activity.

The second form is, then, *contingent areal prediction.* Future centers of volcanic activity are identified in much the same manner that we have identified centers of earthquake activity. The long-range view is represented by maps showing locations of volcanic activity during the past several thousand years.

The short-range view is represented by geophysical and geologic studies of centers that are giving signals of renewed activity. The long-range view of volcanic-hazard identification has been developed very thoroughly by Crandell, Mullineaux, and Miller. Further development of their approach is described in Bailey et al.. Work in Iceland has been described in Thorarinsson and by Sigvaldson.

In Iceland, the mapping of centers of volcanic activity is based on a record from 1100 years of settlement. During this time, eruptions have occurred about every 5 to 10 years. Written records document the location and some of the effects of many of these eruptions.

In the United States, the historical record dates only to the mid-nineteenth century, although this record may be supplemented to a degree by studies of native American settlements. Therefore the bulk of the U.S. record comes from *geological* studies of volcanic deposits.

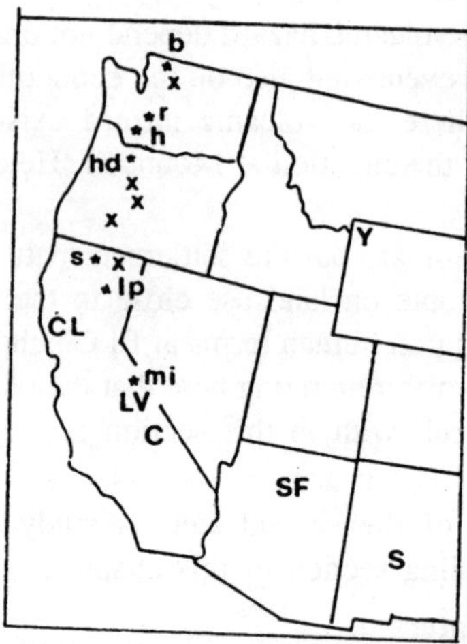

Figure 7.3 : Volcanic hazard map for western conterminous United States. Asterisks mark location of Group I volcanoes active in historic times: b = Mount Baker, r = Mount Rainier, h = Mount St. Helens, hd = Mount Hood, s = Mount Shasta, lp = Lassen Peak, mi = Mono-Inyo craters. Symbol x marks location of Group 2 volcanoes that last erupted more than 1000 years ago. Group 3 volcanoes last erupted more than 10,000 years ago but overlie large magma reservoirs: Y = Yellowstone, S = Socorro, SF = San Francisco Peak, C = Cosco, LV = Long Valley, CL = Clear Lake.

These geological studies are used to create chronologies of those events that have left their mark in the geologic record. There may also be other events that have not left any record.

Information obtained from geologic studies has been presented in maps ranging in scale from a map of the western United States to maps of individual states, and maps of specific target areas like the Long Valley-Mono/Inyo areas in California. The map of the western United States shows the location of three kinds of centers of volcanic activity.

Group 1 includes all volcanoes that have erupted every 100 to 200 years or so, or that have been active in the past 300 years. Mount St. Helens is one such volcano, and Mono-Inyo craters in California are another such source.

Group 2 comprises volcanoes that erupt every 1000 years, or that have not erupted in the past 1000 years. Mount Mazama, the volcano in which Crater Lake in Oregon was formed, is a representative of a Group 2 volcano.

Group 3 comprises volcanoes that have not been active in the past 10,000 years but that overlie large bodies of still molten magma. Long Valley caldera, California, and Socorro, New Mexico, are such areas. Chronologies for many of these centers have been published.

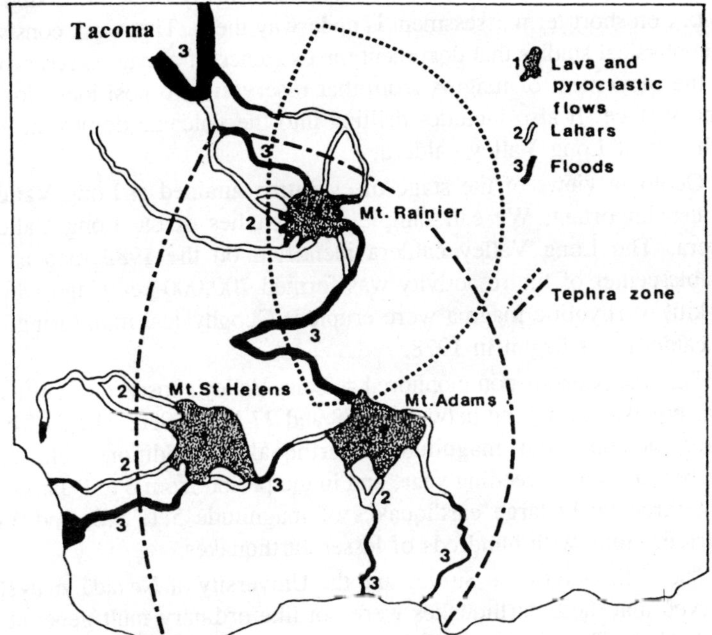

Figure 7.4 : Volcanic hazard map of the southwest part of the state of Washington compiled in 1975. Map shows expected extent of hazard zones around three volcanoes.

The way in which geologic and geophysical studies of one of these centers can lead to more intense investigation is nicely illustrated by the chronology of research on the Long Valley and Mono areas in California. The large-scale map shows three centers that have been identified by the USGS as sources of one or more extremely explosive large-scale eruptions in the past one-million years.

These are Yellowstone caldera in northwest Wyoming, Long Valley caldera in California, and Valles caldera in northern New Mexico. Each of these is a great circular depression tens of km in diameter that formed when an extraordinary volume of rhyolite magma was erupted from a large near-surface reservoir.

At Yellowstone, the volume is estimated to have been 1000 km^3 (compared with 0.6 km^3 at Mount St. Helens); the eruption left the roof of the reservoir unsupported, leading to the collapse now marked by the caldera. All of these centers are under observation. The nature of these observation programs depends on the degree to which geologic and geophysical studies suggest that careful monitoring can be scientifically and economically justified.

The Long Valley area has been deemed worthy of detailed study, so work on short-term assessment is underway there. This work consists of geophysical studies that document the existence of a magma reservoir and the movement of magma from that reservoir into positions closer to the surface. It also includes drilling into the volcanic domes on the north rim of Long Valley caldera.

Geologic views of the stage of evolution attained at Long Valley are also important. We can sample some studies of the Long Valley caldera. The Long Valley caldera identified on the 1983 map as a possible center of future activity was formed 700,000 years ago when 600 km^3 of rhyolitic magma were erupted. Geophysical monitoring of this caldera was begun in 1978.

The first phenomenon monitored at Long Valley consisted of clusters of earthquakes occurring between 1978 and 27 May 1980. The 27 May cluster included four magnitude-6 earthquakes. Additional clusters occurred in each succeeding year, and in the period from 6 to 9 January 1983, three fairly large earthquakes of magnitude 5.1, 5.6, and 5.0 occurred, along with hundreds of lesser earthquakes.

The U.S. Geological Survey and the University of Nevada analysts observed that these earthquakes were not like ordinary fault-generated earthquakes. They suggested that these earthquakes might be generated during the injection of magma. This led to additional monitoring, which included releveling along U.S. Route 395, gravity surveys, magnetic surveys, and continued analysis of the seismic record.

It was found that the ground surface had been uplifted 45 cm in five years. This uplift was occurring above a known magma chamber. The Ryalls found that the earthquake hypocenters had been migrating progressively closer to the surface, indicating that the more recent earthquakes had occurred at depths less than 3 km.

In addition, the Ryalls demonstrated that some of the large earthquakes had been accompanied by a different kind of seismic activity, called *harmonic tremor*. Experience at Mount St. Helens had shown that harmonic tremor indicates that magma is moving upwards.

As a result of these analyses, the Ryalls and Savage and Clark postulated that magma had been rising from the top of a magma reservoir only 8-km deep, causing broad uplift and also leading to the formation of a thin body of magma injected even closer to the surface as a dike. The results of this interpretation are illustrated diagrammatically in a cross section.

All of these findings could be viewed in a context provided by the experience from Mount St. Helens. This led the U.S. Geological Survey in 1982 to issue a warning, which showed the possible centers of eruptive activity and the areal extent of the effects of such eruptions, should they occur.

In 1983, the rate of uplift slowed, and earthquake activity decreased. At the same time, the USGS changed its system for issuing hazard warnings, so that warnings would be issued only if sites such as Long Valley posed an immediate threat. Once centers like Long Valley have been identified and studied, work can be concentrated on more careful prediction of the areas that can be expected to be affected by an eruption.

Predicting the Location of Areas to Be Affected by Eruption

In the second stage of areal prediction, attempts are made to specify the size and location of areas that are likely to be affected by the various forms of eruption products. The previous case studies provided samples of all these eruption products, and they also showed the areal extent of each.

Now we can look at efforts to use the entire geologic and historic record to make these kinds of predictions for many different locations. The prediction of the effects of various kinds of events is well represented by maps prepared by the U.S. Geological Survey. Let us look first at generalised predictions of the areas likely to be affected by tephra falls, lava flows, and pyroclastic surges.

The extent and thickness of a tephra fall depend on the volume of tephra injected into the atmosphere and on the prevailing winds at the time of eruption. We have seen modern examples of this at Mount St. Helens. But well before the 1980 experience, maps of ash fall or tephra-hazard zones for all of the principal sources in the western United States had been prepared.

Figure elsewhere in this chapter is an example of this kind of map, showing areas that would be covered by at least 5 cm of tephra, given two different scales of eruptions. A similar map was prepared for the Long Valley caldera and Mono/Inyo craters in 1982 and was issued as part of the warning.

Tephra flows do not cover areas as large as tephra falls, and we must look at larger-scale maps of smaller areas to see the kinds of predictions that have been made. The volcanic hazard map for the state of Washington shows this very well because it only indicates generalised areas to be affected by eruptions. More detailed maps have been prepared for the areas around each of five volcanoes in which tephra flows, lava flows, and mudflows can be expected. This kind of mapping is based on the geologic record, which shows the limits of deposits of these kinds formed by historic and prehistoric eruptions. An example of such a map is the one prepared for Mount St. Helens.

All of these maps were prepared well in advance of the 1980 eruption of Mount St. Helens, and all were relatively generalised. However, the techniques developed in preparing them proved to be extremely useful. Early events at Mount St. Helens suggested to USGS analysts that there was merit in preparing up-to-date maps showing areas that could be hazardous.

These maps were to be used in preparing for warning and evacuation. One such map, prepared by the USGS on 1 April 1980, showed the predicted extent of effects that would be associated with events of three different scales. This map shows that small- and medium-scale events were expected to produce tephra flows restricted to an area having an 8-km radius around the volcano.

The map also shows that a large event was expected to provide sufficient material and energy to send tephra flows down the various river valleys. Similarly, the larger the event, the further the mudflows would extend down the valleys. Survey analysts also stated that a large slide could be expected on the north side of the volcano, but the extent of the area to be covered was uncertain. The areal prediction shown on this map were borne out by the distribution of the effects of the 1980 eruptions.

COMPREHENSIVE PREDICTION OF VOLCANIC ERUPTIONS

The identification of potential centers of eruptive activity and the mapping of areas that will be covered by some form of volcanic material, by glacier meltwater, or by lahars and avalanches, all help to establish how severe the hazard at a particular center may be. Where the hazard is severe, there is a powerful motivation to work on the comprehensive prediction of volcanic eruption.

Quite simply, predicting the time of eruption, or specifying a limited time interval within which eruption is highly probable, is fundamental to developing warning and evacuation systems. The ability to do this has been developed with considerable success for eruptions of Mauna Loa and Kilauea on the island of Hawaii.

It has been much more difficult to do this at other locations such as the Cascades, the Caribbean, and Central America. It appears, in

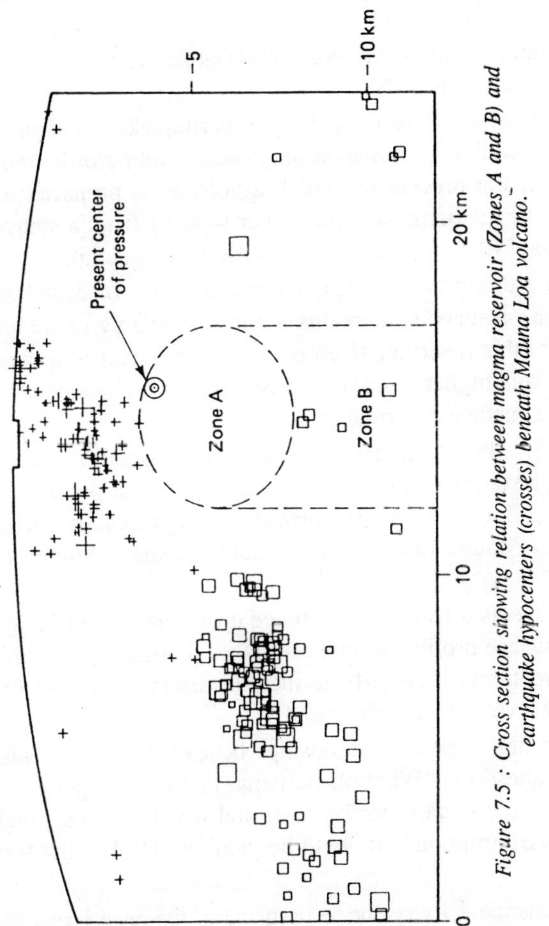

Figure 7.5 : Cross section showing relation between magma reservoir (Zones A and B) and earthquake hypocenters (crosses) beneath Mauna Loa volcano.

fact, that the difficulty of making a comprehensive prediction increases with the severity and danger of the event to be predicted. Thus Hawaiian eruptions, which *can* be reliably predicted, are not particularly dangerous; Caribbean or Central American eruptions, which *cannot* be reliably predicted, are very dangerous.

We will examine these differing capacities after first presenting the basic situation that precedes most, if not all, volcanic eruptions, using it as a guide to the search for precursors usable for temporal prediction.

Most volcanic eruptions are preceded by near-surface intrusion of magma from deep sources. These bodies of magma serve as high-level reservoirs. An eruption taps these reservoirs and drains and depressurizes them. One or more of the reservoirs must then be repressurised before further eruptions can recur. We can visualize this by using cross sections for the Hawaiian Islands.

Decker et al. have used data on earthquakes beneath Mauna Loa volcano, as well as information on ground-level deformation, to develop a picture of the process by which a volcano is prepared for eruption. They believe that magma is fed rather steadily from a source at a depth greater than 40 km to a reservoir at a 10-km depth.

During this process, many earthquakes are recorded at a depth of 40 km. The reservoir maintains a constant supply of magma. Magma is fed from this reservoir to an overlying reservoir at a depth of 5 km. Filling of this higher reservoir causes shallow earthquakes and readily observable surface deformation.

In 1983, these analysts observed that no precise forecast of future eruption of Mauna Loa could be made, but "if the present rate of strain continues to increase . . . the probability significantly increases for an eruption of Mauna Loa during the next 2 years". Mauna Loa erupted in 1984.

We have as a frame of reference cross sections of Kilauea volcano showing surface profiles at different stages. Beneath these surface profiles are crosssectional views of the magma reservoir and feeder pipes. A deeper reservoir is also indicated.

Given this view of relationships beneath volcanoes, analysts have asked the question, "What phenomena might accompany the transition from one of these states to the next, and which of these might serve as precursors of eruption that could be monitored?" We list some of these phenomena:

1. Seismic activity due to fissuring of the deep rocks and displacement of these rocks past one another.
2. Seismic activity generated along the walls of the feeder pipes as magma moves upward through them.
3. Uplift of the surface over the reservoir.
4. Changes in the transmission of seismic waves along paths that cross the magma reservoirs.
5. Emission of gases, changes in surface temperature, formation of fumaroles, changes in magnetic properties, etc.

It has been shown at the Hawaii Volcano Observatory (HVO) run by the U.S. Geological Survey that several of these phenomena serve as precursors of eruption, allowing reasonably specific forecasts to be

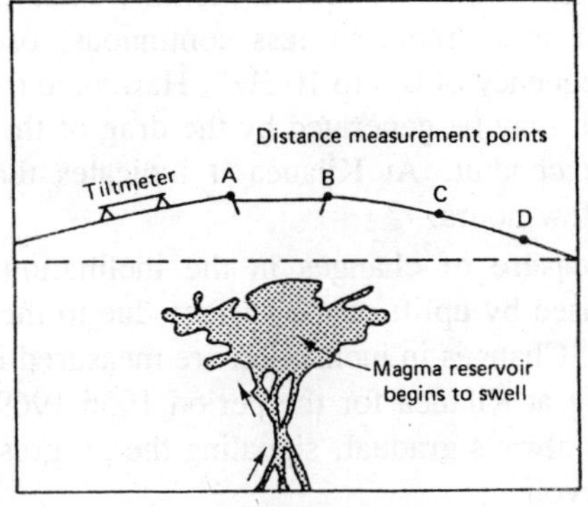

Stage 1
Inflation begins

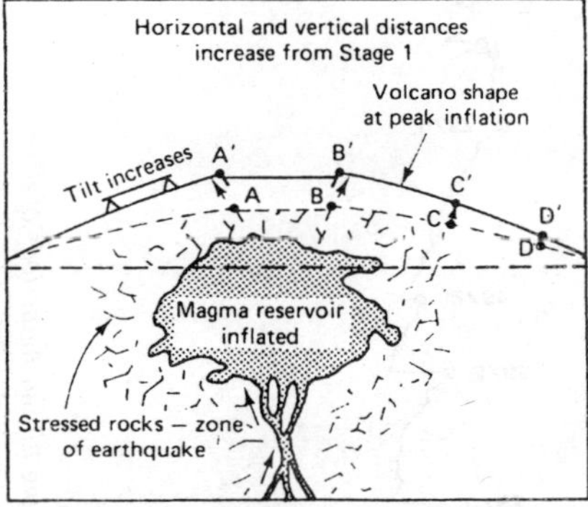

Stage 2
Inflation at peak

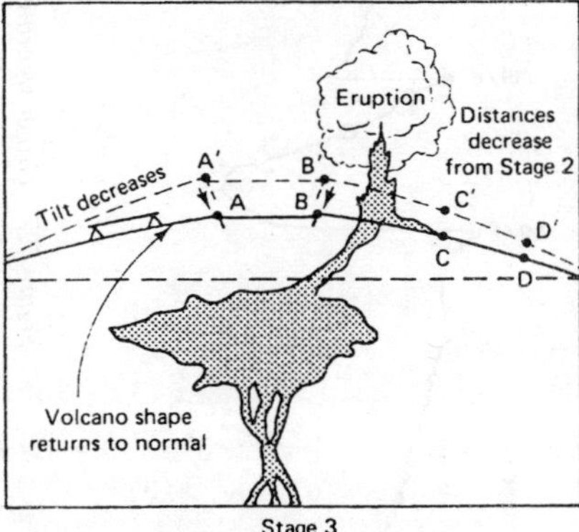

Stage 3

Figure 7.6 : Diagrams showing stages in the course of a typical eruption.

issued. The two most reliable are *harmonic tremor* and *surface tilt*. Harmonic tremor is a "more or less continuous, oscillating ground motion with a frequency of 0.5 to 10 Hz". Harmonic tremor is not well understood, but it may be generated by the drag of the magma against the walls of the conduit. At Kilauea it indicates that eruption will follow within a few hours.

Tilt is a measure of changes in the inclination of the ground surface. It is caused by uplift of the surface due to the inflation of the magma chamber. Changes in inclination are measured in microradians. A graph of tilting at Kilauea for the period 1956-1969 shows that the increase in tilt is always gradual, signaling the progress of inflation of the magma reservoir.

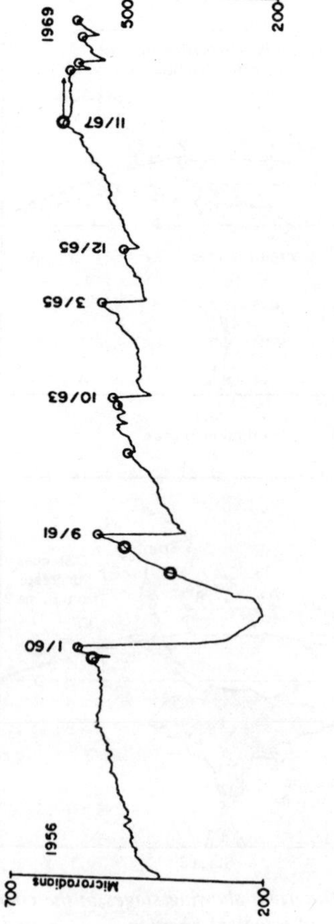

Figure 7.7 : Graph showing increasing tilt during periods of inflation and sudden change at times of eruption.

The end of this process is signaled by an abrupt decrease in tilt value, which results from the eruption. Every period of increasing tilt terminates with an eruption. It is evident from the graph that eruption becomes more and more likely as the tilt value approaches 600 microradians.

Therefore monitoring tilt can lead to more and more narrowly bracketed estimates of the time interval in which an eruption is to occur. This method has been refined so that the Hawaii Volcano Observatory is able to issue quantitative 1-day, 7-day, and 30-day forecasts of the probability of eruption of Kilauea.

These eruptions are nonviolent, and they typically produce lava flows. The changes preceding them can be monitored safely, and forecasts can be issued regularly. Now let us compare this situation with the situation at Mount St. Helens.

It appears at Mount St. Helens that the three most reliable precursors were the first three in our list: (1) increased seismic activity, (2) harmonic tremor, and (3) uplift.

At Mount St. Helens, uplift and outward movement of the north side of the volcano were a clear and spectacular indication of the accumulation of magma in a near-surface reservoir (Fig. 10.22). This reservoir is shown in cross section as *a cryptodome.* All analysts understood that the volcano could not continue to expand at the observed rate without an event taking place. This provided a basis for predicting an event on a time scale measured in weeks.

In addition, although it was not fully appreciated at the time, the 60 days of virtually continuous major seismic activity at depths averaging 3.3 km below mean sea level indicted an imminent eruption. The synthesis of observations on uplift and of the steady occurrence of near-surface earthquakes led the geologists to state that activity was so likely that the state of Washington might wish to issue and enforce a warning, which it did.

Following the 18 May 1980 eruption, five additional major eruptions occurred in 1980. Study of the patterns of harmonic tremor preceding each of these and of the shallow earthquake activity enabled the analysts to develop their prediction method so that the last four of this group of five were all predicted successfully. This experience led in turn to considerable success in predicting a major eruption that occurred on 19 March 1982.

The predictions issued for the 19 March 1982 eruption illustrate very nicely how a prediction can be made more specific with time. In

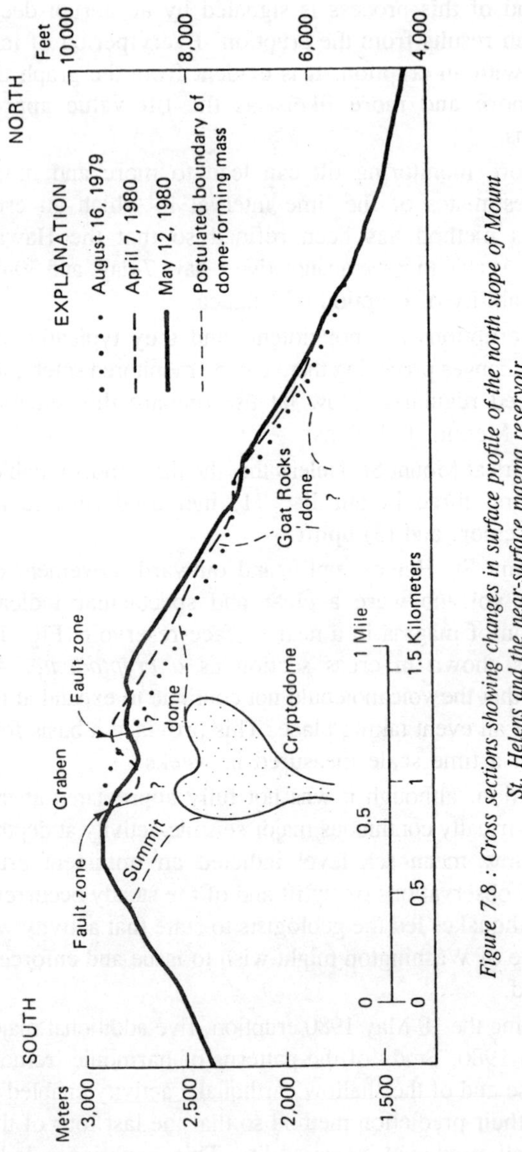

Figure 7.8 : Cross sections showing changes in surface profile of the north slope of Mount St. Helens and the near-surface magma reservoir.

March 1982, the pattern of earthquakes and summit uplift led to the first prediction issued on 12 March 1982, stating that an eruption could be expected within three weeks of that date. On 15 March, a second prediction was issued, stating that an eruption could be expected within I to 5 days. Then, on 19 March at 09:00, an alert was issued that an eruption could be expected within 24 hours. The eruption occurred at 21:30.

These findings on Hawaii and at Mount St. Helens show that the ability to predict is improving with experience. This ability is now being employed in the monitoring of the Long Valley caldera. Let us now examine the use of this ability in adapting to volcanic hazard.

ADAPTATIONS TO VOLCANIC HAZARD

Human settlements have always encroached on volcanic hazard zones because there were benefits to be obtained from doing so, or because the nature of the hazard was not appreciated. Some of these settlements have been affected by eruptions, but they usually have been able to deal with them acceptably.

Others have been subjected to extraordinarily destructive events, such as those recorded at Pompeii, Santorini (Thera) in the Aegean Sea, Martinique in the West Indies, and Tambura and Krakatau in the South Pacific. Descriptions of these are in Sheets and Grayson (1979), Tazieff and Sabroux (1983), and in the hundredth anniversary volume on Krakatau (Simkin and Fiske, 1983).

In the case of events as destructive as the eruptions of Tambura and Krakatau, the only adaptations that could have been employed were total avoidance of the areas or successful prediction and evacuation. Let us sample the adaptations and the problems met in employing them.

The same adjustments are theoretically available to those who face volcanic hazard as to those who face the other hazards dealt with in the preceding chapters. But the nature of volcanic hazard seems to militate against employing many of these adjustments.

Volcanic events are so rare that avoidance of use of land within zones that might be affected is not often seen as feasible. Yet these rare events are potentially so destructive that the search for usable adjustments seems to be worth continuing.

Experience in regions where lava flows are the principal threat shows that it is always possible to evacuate human settlements before they are inundated. However, it is not possible to save the structures. Therefore the question for those who would settle in such areas is to decide whether the assurance that evacuation is feasible is sufficient to justify taking the risk of building and property losses that can only be handled through insurance, if at all.

Hodge et al. examined these adaptations on Hawaii and found that people were prepared to take the risks. In all other situations, it appears that short-term adaptations such as evacuation, temporary restrictions on land use, the drawing down of reservoir levels, and the preparation of emergency systems are the principal means for dealing with volcanic hazard, so it is worth looking at their employment.

To do this we will draw largely on experience at Mount St. Helens as reported in Foxworthy and Hill and in the following sources in Lipman and Mullineaux.

Short-term adaptations at Mount St. Helens were almost all the result of decisions by public officials to take action. In each instance, these officials were dependent on the scientific advisory group, which had to present its predictions and estimates of risks on a day-to-day basis.

We have seen that one of the major presentations of this kind of information took place on 30 April 1980 when the up-todate hazard evaluation map was presented to public officials. At that time, officials from the state of Washington and from the U.S. Forest Service established two categories of restricted-entry zones around the volcano. A Red Zone was outlined from which all but scientific observers and emergency personnel were banned.

A Blue Zone was established outside of this within which logging operations were allowed to continue and in which property owners could be present during daytime hours only. Foxworthy and Hill explained how these zones were progressively enlarged from the time that they were first put into effect through a period following the 18 May eruption.

The critical decision to enlarge these zones was made on 30 April 1980, at which time the Red Zone was extended further north to include all of the area in the National Forest surrounding Spirit Lake. We can appreciate the importance of this decision by examining a map that shows the boundaries of the Red Zone established on 30 April, the position of observers officially allowed to remain within the zone, and the position of other eyewitnesses.

We can appreciate the effects of this by reviewing the USGS record compiled on the day of the eruption, 18 May 1980. The locations of these reports are shown on the map. The first to radio a report on the 18th was geologist David Johnston, who only had time to say, "Vancouver, Vancouver, this is it," before the directed blast overwhelmed his station.

The next warning came from observer Gerald Martin, who radioed, "The camper and the car just over to the south of me are covered. It (the blast cloud) is going to get me, too." His station then was inundated by pyroclastic surge material. At Bear Meadow, 11 miles northeast of the summit, just outside the Red Zone, photographer Gary Rosenquist was able to take a series of photographs, awaken other campers, and escape by driving north.

Keith Ronnholm, another photographer at the same campsite, delayed his departure a few seconds longer, and escaped only by driving through the tephra falls that lowered visibility to zero. To the northwest along the North Fork of the Toutle River at a point west of the Red Zone, three loggers were inundated by a very hot black cloud that burned them and buried everything under a foot of gray ash.

The three men died after being rescued. The U.S. Forest Service estimated that at least 2000 lives had been saved by the decision to create the Red Zone. The manager of Spirit Lake National Forest area observed that on a normal weekend at that time of year at least 2000 people would be camped there.

He also observed that the publicity generated by the volcano would presumably have attracted hundreds or thousands more. As a result of this decision only 60 lives were lost. Examination of the map shows that more lives could have been saved if the western boundary of the Red Zone had been drawn to reflect the geologists' predictions rather than political boundaries.

The west boundary of the Red Zone was a county boundary, cutting neatly across a high-hazard zone. Mudflows and flood water moved in all directions. The flows that moved down Pine Creek destroyed Eagle Creek Bridge at 09:00 and raised the water level by 2 m in 15 minutes at Swift Reservoir. The reservoir level had been lowered, so the dam was not overtopped. Mudflows moving down the Toutle River drainage took longer.

At 10:15, the mudflows reached the junction of North Fork and Smith Creek as a 4-m high wall of water. Following the 18 May 1980 eruption, the Red and Blue zones were enlarged. At the same time, newly created hazardous situations had to be assessed. For example, the debris avalanche had blocked Spirit Lake, so that the lake level had risen 60 m.

Youd et al. analyzed the stability of this new dam and concluded that it was stable, providing assurance that the new Spirit Lake did not constitute a new source of hazard. Jennings et al. analyzed the flooding that would result as various dams created by mudflows on the North Fork of the Toutle River failed. These provided the basis for decisions concerning monitoring of these dams and control of land use in these flood-hazard areas.

All decisions to restrict access and land use generate substantial protests as shown at Mount Baker in 1975, at Soufriere in the Caribbean in 1979, and at Mount St. Helens. Many of these protests have economic roots, as exemplified by the arguments put forth by loggers at Mount St. Helens, saying that preventing them from having access to the Red Zone would cause economic hardship for them.

This was also illustrated at Mount Baker, where decisions in March of 1975 by the U.S. Forest Service to control access meant the loss of income from the principal resort owner there. These protests also are rooted in beliefs that the predictions of the scientific group are not credible. Many examples of this are to be found in the Foxworthy and Hill report.

Decisions to order evacuations of inhabited areas are even more difficult since they can be very costly. If the predicted eruption fails to take place, then scientists and public officials are placed in a very difficult position. This was shown at Soufriere in the Caribbean April 1979, when the threat of eruption led to heated debate concerning the desirability and feasibility of evacuation.

Eruptions did occur and an evacuation of the northern part of the island was ordered in late April 1979. The USGS volcanic hazards group has concluded from all of its experience that the best approach is to work to develop the ability to predict effects within the range of the probable and to find ways to make potential users understand the limits of confidence with which each prediction or forecast is expressed.

8

The Coast and Coastal Management

More than half the people in the United States live close to the coast, and many are willing to pay dearly to live at the water's edge. Homes, apartments, hotels, and motels crowd many segments of the shore, competing with other uses of coastal land-public recreation, power generation, refineries, and military bases. This competition makes coastal land especially valuable.

Historically, most of this land (in the United States) has been allocated by "allow(ing) supply and demand to determine the usage of coastal areas through the price mechanism—the use which paid the highest price for a particular property obtained . . . .

This market-allocation system has worked well for land away from the coast, but it has been challenged by those who believe that coastal land poses special problems that demand a different system of allocation. What are the special features of coastal land that give rise to these problems?

1. The boundaries between coastal land, wetlands, and the water shift constantly, making the coast a high-hazard zone.
2. Coastal wetlands and near-shore lake and ocean waters are biologically productive and ecologically and economically important.
3. Wetlands and near-shore waters are easily harmed by human modification of the coast and by disposal of wastes.
4. Coastal land and water are highly valued for their recreational

potential.

5. Major petroleum and other resources are found along many coasts.

Each of these features provides a focus for those who argue that market allocation of coastal land and offshore resources has not and cannot take into account the negative effects that unregulated development has on the coastal zone.

Those who make this case include groups of coastal-zone and marine geologists and biologists, conservationists, and supporters of increased public access to the coast.

From time to time these groups have successfully put their views before the public and the policy-makers, and by so doing, have stimulated analysis and change. Here is a sample written by a geologist concerned with shoreline erosion: Pilkey's "*Truths of the Shoreline*"

I. There is no shoreline erosion problem until someone builds something on the beach to measure it by.

II. Construction on the beach reduces flexibility and in itself causes erosion.

III. The interests of beach property owners should not be confused with the natural interest.

IV. Once you start stabilisation, you can't stop.

V. The cost of saving beach property is, in the long run, greater than the value of the property to be saved.

VI. In order to save the beach, you destroy it.

These statements reflect a point of view that is clear and controversial. This concluding chapter is a good place to consider such controversial positions because they reflect the reality of the science, practice, and policy of land and water resources and crop up often when people confront the management of the coast.

We cannot examine all facets of land and water use in the coastal zone here, but we can sample a variety of studies to show how scientific findings and scientists influence coastal practices and policies.

The first part of this chapter presents information about the coastal zone and the physical and biological interactions there. Readers familiar with this material will want to go directly to the discussion of longstanding and still-developing issues.

These include: shoreline erosion, the appropriate use of freshwater and saltwater wetlands, and the management of estuarine systems.

THE COPASTAL ZON

Inman and Nordstrom have provided a physical framework and a nomenclature that are widely used in discussing the *coastal zone*. The coastal zone extends from the edge of the continental shelf inland across the coastal plain as shown in the figure. Inman and Nordstrom distinguished two classes of coast:

(a) coasts with barrier islands, and

(b) coasts with sea cliffs. Coasts fringed by coral reefs are a third important type. There are also the inland coasts of the Great Lakes, which display features analogous to those shown in Figure elsewhere in this chapter.

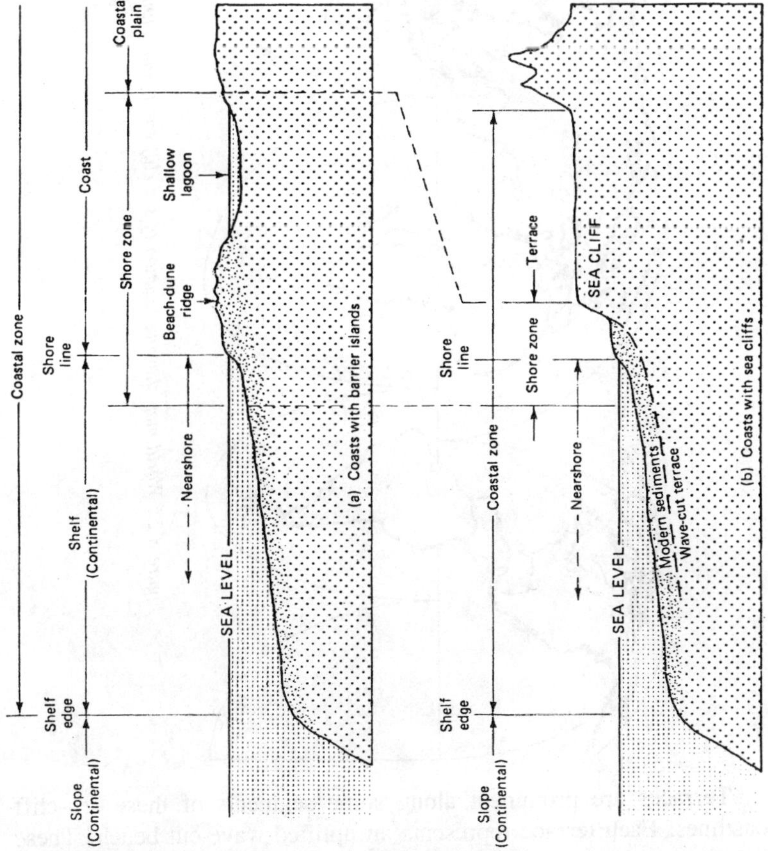

Figure 8.1 : Features of the coastal zone as seen on the Atlantic and Gulf coasts (a) and the Pacific Coast of the United States (b).

SEA-CLIFF COASTS

The most common type of coast, worldwide, is the sea-cliff coast. Emery and Kuhn have shown that sea cliffs are present along approximately 80 percent of ocean coasts. The continental shelf along this kind of coast is narrow.

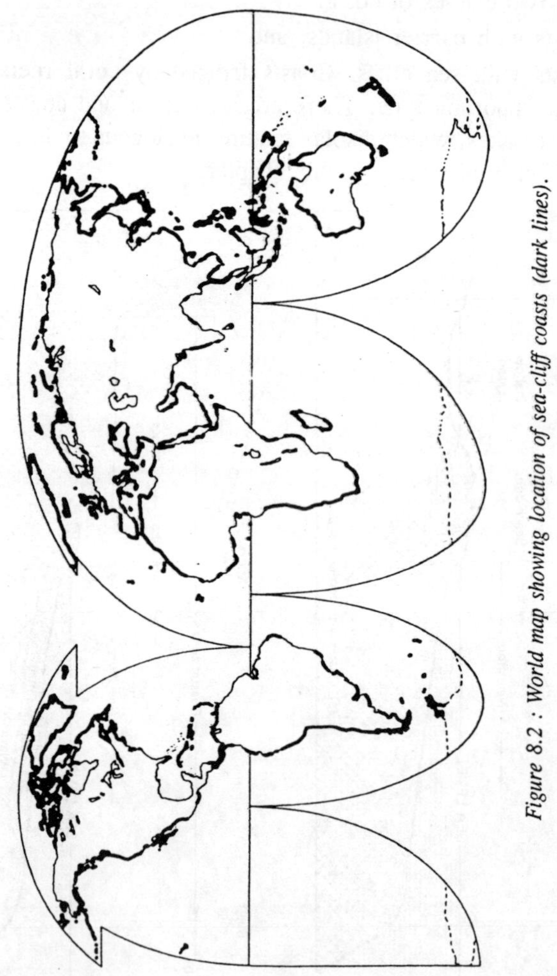

Figure 8.2 : World map showing location of sea-cliff coasts (dark lines).

Terraces are prominent along some segments of these sea-cliff coastlines. Each terrace represents an uplifted wave-cut bench. These cliffed coasts tend not to have barrier islands and wetlands. Emery and Kuhn list the following three types of sea cliffs:

1. Active sea cliffs are those that are being actively eroded by wave action. Debris that falls to the base of the active sea cliff is rapidly removed by wave action.
2. Inactive sea cliffs are characterised by the accumulation of talus or debris at the base of the cliff.
3. Former sea cliffs are hidden beneath accumulated debris.

Emery and Kuhn argue that understanding of the mechanisms and rates of retreat of sea cliffs is needed because more sea-cliff areas will be considered for development. They observe that people who consider sea-cliff terrain for residential development may believe that the terrain is more stable than it really is.

The reason for this misconception is that sea cliffs retreat in bursts of activity separated by longer periods with no retreat. This view is important in considering the Pacific Coast and the Great Lakes.

Barrier Islands

Approximately 2400 km of the Gulf and Atlantic coasts of the United States are bordered by barrier islands; this involves nearly half the total length of this coastline. These islands separate the open ocean from the shallow lagoons that lie between island and mainland.

Geologic studies have established that these islands have been built up from sand over the past 10,000 years or so, when the sea level was rising. A great many of them have been built in just the last 1000 years. We can characterize these islands in terms of the forms they display

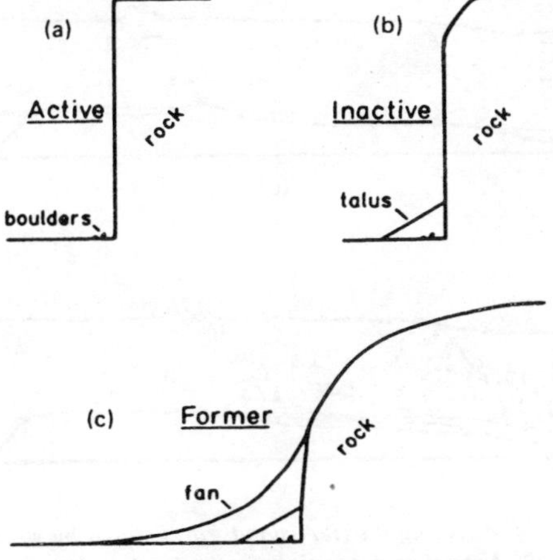

Figure 8.3 : Cross sections of sea cliffs at three stages in their development.

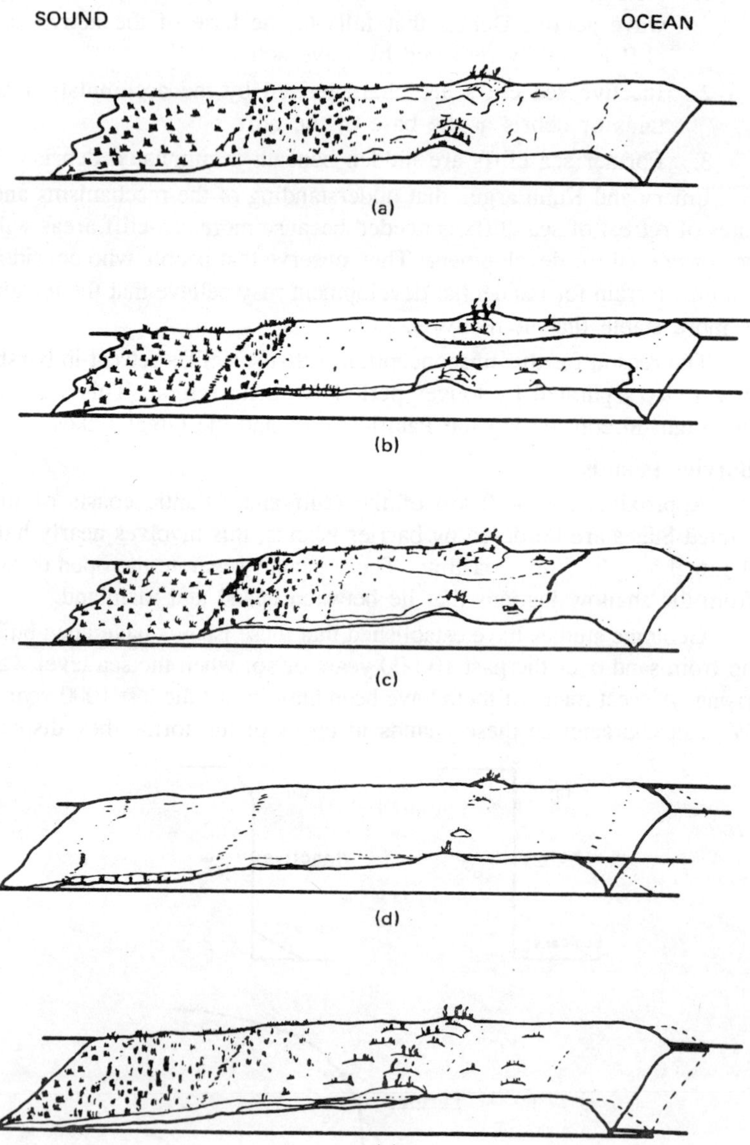

Figure 8.4 : Segment of migrating barrier island (a) modified by storm-driven erosion and overwash (b, c, d). Note that deposition on the landward (sound) side builds the island up and allows it to migrate landward. Note also that the form of the island in (e) is similar to the form of the island in (a), but the position of the island has shifted west.

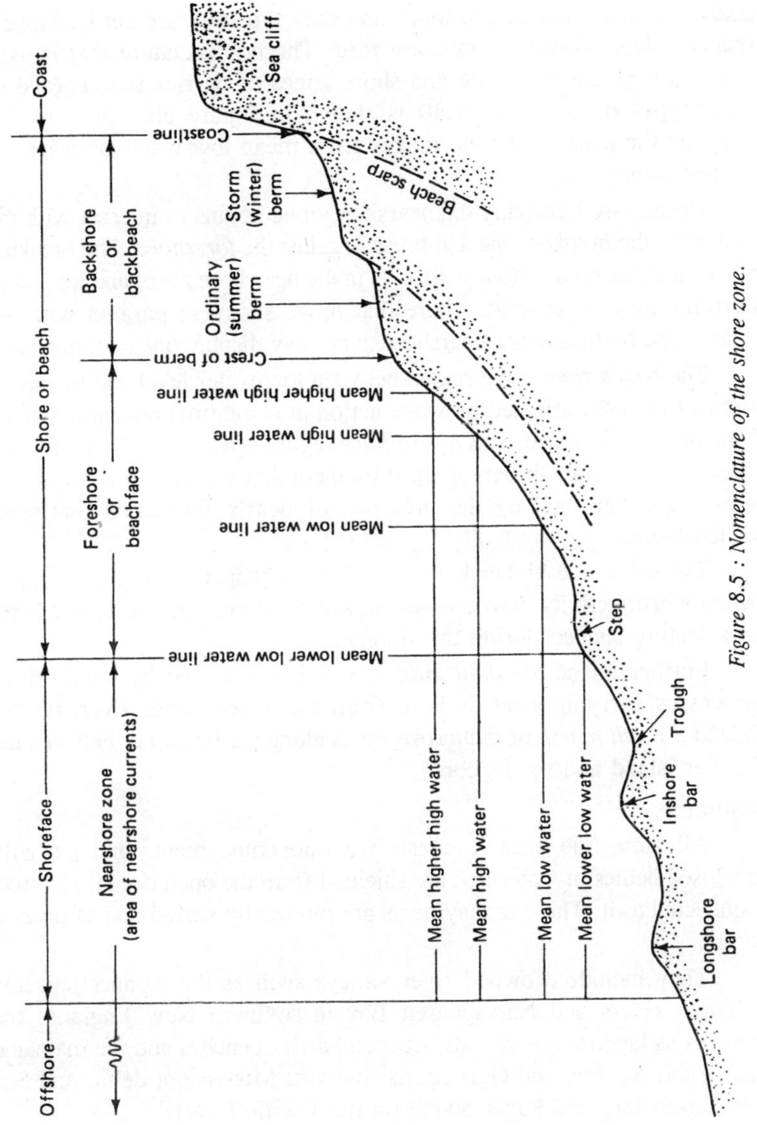

Figure 8.5 : Nomenclature of the shore zone.

in plan view and in cross section. The plan view of barrier islands depends on the conditions under which they have evolved. For example, along coasts where the tidal range is small (1 to 2 m), barrier islands tend to be tens of km long, are cut by very few tidal inlets, and are overlain by many fan-shaped deposits of sand called *washover fans*.

Along coasts where the tidal range is greater (2 to 4 m), the islands tend to be less than 20 km long, and they typically are cut by closely spaced inlets. Washover fans are rare. The nomenclature that is used in discussing the nearshore and shore zones of barrier islands (and of other types of coasts as well) is shown in Figure elsewhere in this chapter: the *nearshore zone* ranges from mean low water outwards to deeper water.

Ocean swell entering the nearshore zone begins to interact with the bottom in the *breaker zone* and breaks against the *foreshore*. The breaking waves interact with bottom sediments in the nearshore zone, and longshore currents may be generated here that move sediment parallel with the coast. The bottom in the nearshore zone may display one or more bars.

The beach zone is the region between low-water level and the upper limit of the area affected by wave action at some time during the year. This beach zone consists of *a foreshore* region affected by the swash and backwash of water driven up on it by incoming waves and a *backshore* region distinguished by the presence of nearly horizontal platforms called *berms*.

Typically, the higher berm is a feature shaped during the winter, when storm activity drives water higher onshore, and the lower berm is a feature shaped during the summer.

Further inland are *dune ridges*, which are shaped by wind-driven processes carrying sand onshore from the beach zone. Even further inland are *salt marsh* or *mangrove* zones along the boundary between the barrier island and the lagoon.

Estuaries

All along the oceanic coasts we find embayments and partially enclosed bodies of water that are shielded from the open ocean, although connected to it. These embayments are physically varied and of diverse origin.

They include drowned river valleys such as the Connecticut and Thames rivers and Narragansett Bay in southern New England, the numerous lagoons and sounds between barrier beaches and the mainland along the Atlantic and Gulf coasts, the vast Mississippi delta, and San Francisco Bay and Puget Sound on the Pacific Coast.

Each of these embayments is a transition zone between land and sea in which fresh water and sea water mix and in which there is extensive biological exchange. These embayments are called *estuaries*.

Estuarine ecosystems have special biological importance because they are regions of high primary productivity as reflected in a very high

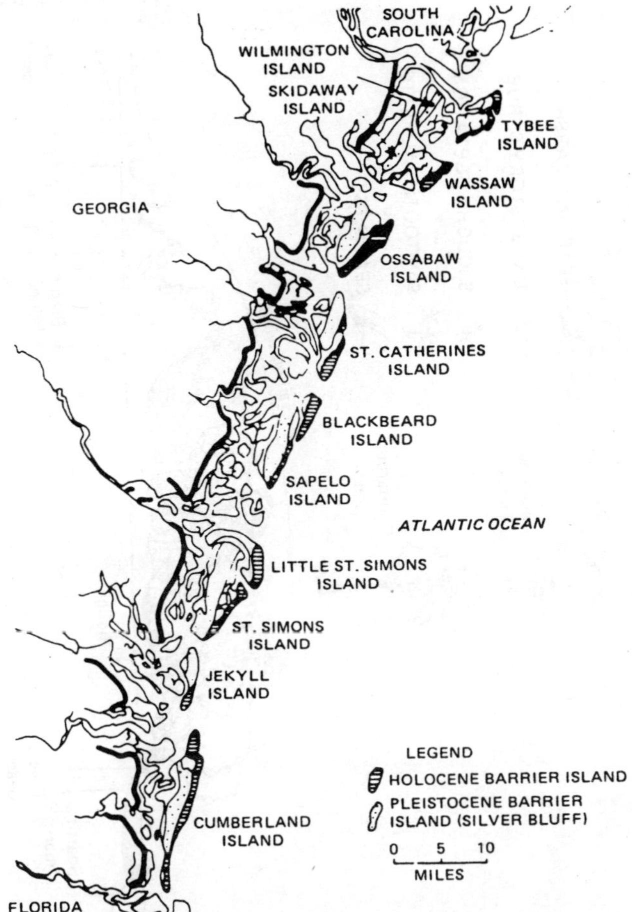

Figure 8.6 : Barrier islands and estuaries of the Georgia coast.

rate of production of plant material and in their role as a nursery and feeding ground for species that spend most of their life cycle elsewhere.

The plant material produced in the estuaries provides the base of important coastal food chains that include *consumers* or plant-eating animals like zooplankton, *foragers* that feed on consumers, and *predators* that feed on foragers or one another.

Each unit of an estuarine ecosystem is characterised by its physical features, by processes that take place within this physical setting, by *modulators* that control these processes, and by biological and visual characteristics.

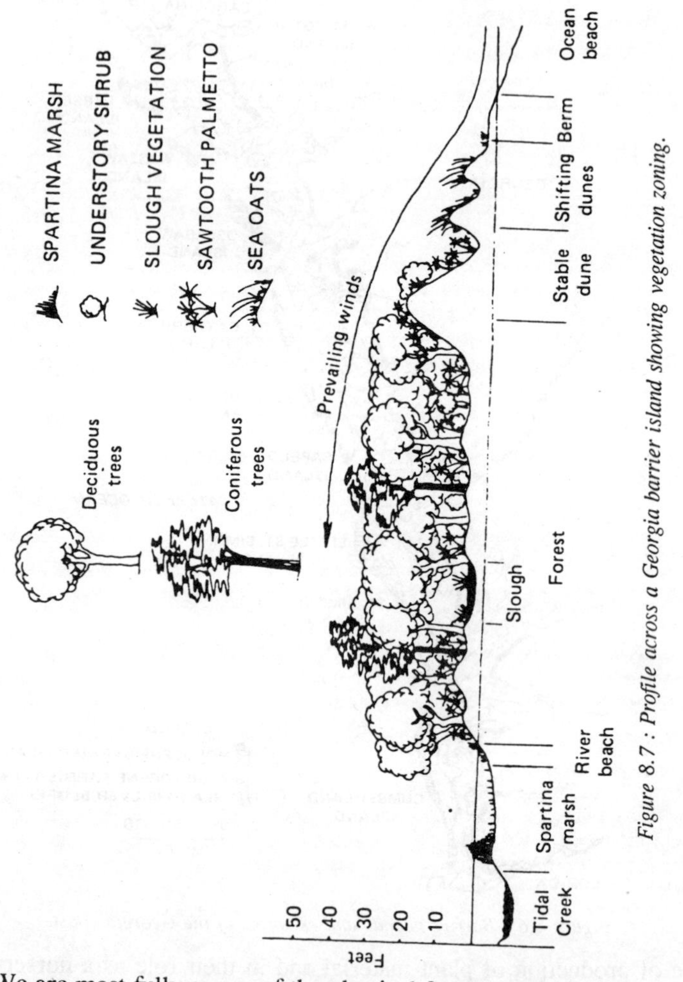

Figure 8.7 : Profile across a Georgia barrier island showing vegetation zoning.

We are most fully aware of the physical features of the air-land and water-land interfaces and of the biological and visual features of the air-land interface. However, we also need to recognize that these features depend on energy exchanges that are steadily taking place in estuarine waters and on the adjacent land. These exchanges depend on fluctuations in salinity, water level, and water chemistry that have both natural and human sources.

We can get some sense of these relationships by looking at a sample estuarine system associated with barrier islands. This will provide us with a more complete picture of the biological features of interest.

Barrier-island Lagoons
An Example from Georgia

Barrier islands and associated lagoons illustrate the interactions between physical and biological processes that are a part of *all* estuarine systems. These interactions must be understood in order to understand coastal evolution and to offer forecasts of the future.

Profiles across barrier islands and the associated lagoons show a range of habitats at the interface between land and air and land and water. Each habitat displays its own biological character as shown by a sample from Georgia.

The Georgia coast is made up of barrier-island estuary complexes, one of which is Sapelo Island, the site of a major coastal research station. A profile across one of these complexes shows the relationship between barrierisland topography and plant zonation.

Toward the open ocean, the outer dune ridges are partially stabilised by sea oats, a plant that can tolerate salt spray from the ocean. Inland of these ridges is a zone of higher and more stable dunes on which palmettos have established themselves. Beyond this is low-lying terrain on which coniferous and deciduous trees are established.

These trees cannot tolerate salt water. The inland side of the island is the edge of a lagoon characterised by salt marsh, mud flats, and tidal channels.

The salt marshes are tidal areas in which a well-developed plant zonation has evolved in response to tidal fluctuations and the resulting variability in water depth, salinity, and nutrient availability. The zonation reflects the distribution of species and also the differences in forms of single species. For a Georgia salt marsh, the zones are as follows:

1. Smooth cordgrass *(Spartina alterniflora)* dominates the intertidal zone. In the range below mean high-tide level, this plant displays a tall form; above this level it displays a short form.
2. At slightly higher levels other species such as needle rush *(Juncus roemerianus)* and salt grass *(Distichlis spicata)* become dominant. These have substantial tolerance for salinity.
3. At elevations at which salinity becomes quite low, shrubs such as the genera *Iva, Baccahris,* and *Borrichia* appear.

Salt-Marsh Ecosystems

The Georgia example is one of many we might have chosen to illustrate features of barrier islands and the associated salt-marsh wetlands. The pattern seen in the saltmarsh wetlands is similar from region to region, but the species that contribute to the pattern differ. In southern Florida and in the Virgin Islands, and in fact in all regions between the 25th parallels north and south, this niche is occupied by mangrove swamps.

In all of these settings, wetlands may be divided into *lower wetlands* and *upper wetlands*. The lower wetlands are all those areas having a surface elevation between mean low tide and mean high tide. They are covered by salt water at any high tide reaching at least the level of mean high tide.

The lower wetlands are marked by *Spartina* or by mangroves *Rhizopora*. The upper wetlands are all areas marked by salt-tolerant species in a zone at elevations ranging from mean high water up to annual high water.

Comparable zonal patterns are represented below mean sea level. These are marked by the distribution of shellfish species, by beds of marine grasses, and by beds of kelp.

Salt marshes are of special interest because they are highly productive. Anyone interested in learning about salt marshes may find it useful to begin by reading the *Life and Death of a Salt Marsh*. Much of what follows is based on this book.

Spartina and algae constitute major *primary producers* in salt marshes. Plant fragments from *Spartina* and other species are distributed in the water, where they are used as food sources by the bacteria.

The bacteria decompose the plant fragments, converting them to a food source for other organisms. These other organisms include larvae of insects, the insects themselves, and a variety of shellfish, crustaceans, and fish.

Many of these species then serve as a food source for other species at higher trophic levels. The result of this system is that the salt marsh produces more organic material per unit area than most of the richest farmland. As the Teals explain, this productivity is possible, because:

1. Nutrients are constantly resupplied due to the tidal ebb and flow through the salt marsh.
2. Waste products are removed from the marsh by this ebb and flow.
3. Cycling of nutrients through plants takes place rapidly and continuously.
4. Production continues throughout the year, even in the north.

Salt marshes and associated estuaries are extremely important as nurseries for many species that spend much of their time in the open ocean. This may be appreciated by tracing the life cycle of any one of a number of species. The National Estuarine Survey did this for the Gulf Coast shrimp.

The Gulf Coast shrimp spawn offshore, but the larvae drift shoreward, and they then take up residence in lagoons. They grow rapidly in these lagoons during a period ranging from 2 to 4 months. They then return to the Gulf to mature fully.

Salt-marsh and other wetlands also serve other functions. They trap fine sediment and by so doing, can grow laterally or vertically to keep pace with the rise of sea level or the lowering of the land due to subsidence.

Marshes also serve as a buffer between human activities and the open water. This buffer is physical, chemical, and aesthetic. The buffer puts distance between people and flood waters and can purify chemically contaminated waters. The marsh also provides a landscape highly valued by many people.

Other Coastal Zones

The Virgin Islands, parts of Florida, and the Hawaiian Islands all display coastalzone segments distinguished by fringing coral reefs. These reefs create lagoons on the landward side that are physically protected from the open ocean but are in free communication with it.

These reefs and lagoons are particularly sensitive to human activities. The Great Lakes shorelines match most of the features of the oceanic coasts. The major differences are the absence of tides and salt water.

Major fluctuations in water level create some of the same kinds of features that tidal fluctuations create on oceanic coasts.

PROBLEMS OF THE COASTAL ZONE

The coastal zone is the site of many conflicts that arise because of the large number of possible uses focused on a narrow strip of land and water. These conflicts include disputes about coastal erosion and the control of erosion; the siting of homes, power plants, and industry; and the management of saltwater and freshwater wetlands. We will sample some of these disputes.

Coastal Erosion

The U.S. Army Corps of Engineers and many geologists and other coastal observers agree that coastal erosion is a major problem in the United States. A 1971 national shoreline study by the U.S. Army Corps of Engineers showed that 20,500 miles of the total U.S. coastline of 84,240 miles was undergoing significant erosion. Some 2700 miles of shoreline were described as being affected by critical erosion.

Many geologists whose research has concentrated on the coastal zone concurred in this estimate. In 1981, 80 geologists who attended a Conference on America's Eroding Shoreline issued a report: *Saving the American Beach: A Position Paper by Concerned Coastal Geologists.* Their position is summarised in the following statements:

1. People are directly responsible for the "erosion problem" by constructing buildings near the beach. For practical purposes there is no erosion problem where there are no buildings or farms.
2. Fixed shoreline structures (breakwaters, groins, sea walls, etc.) can be successful in prolonging the life of beach buildings. However, they almost always accelerate the natural rate of beach erosion. Resulting degradation of the beach may occur in the immediate vicinity of structures, or it may occur along adjacent shorelines sometimes miles away.
3. Most shoreline stabilisation projects protect property, not beaches. The protected property belongs to a few individuals relative to the number of Americans who use the beaches. If left alone, beaches will always be present, even if they are moving landward.
4. The cost of saving beach property by stabilisation is very high. Often it is greater than the value of the property to be saved, especially if long-range costs are considered.
5. Shoreline stabilisation in the long run (10 to 100 years) usually results in severe degradation or a total loss of a valuable natural resource, the open ocean beach.
6. Historical data show that shoreline stabilisation is irreversible. Once a beach has been stabilised, it will almost always remain in a stabilised state at increasing cost to the taxpayer.

Coastal landowners and users have also expressed concern with losses caused by coastal erosion. In a great many instances, they have successfully persuaded policy-makers to pass legislation to provide public funds for the construction of

shore-protection measures and for sand supplies to eroding beaches. These practices are the ones criticised by the 80 coastal geologists cited above.

In short, coastal erosion is an endlessly recurring problem, control of erosion costs money, and the science, engineering, and politics of coastal erosion are all foci of lively disagreement. As a result, coastal erosion maintains a high visibility among both the public and the experts.

Factors Controlling Coastal Erosion

The forms and distribution of beaches, bluffs, and dunes are developed through processes in which there is a tendency to reach an equilibrium determined by environmental conditions. If these conditions are not changed, then the equilibrium forms may not change much with the passage of time.

The shape of Half Moon Bay actually is believed to be such an equilibrium form. When conditions are changed in any way, we can expect the form and location of coastal features to change also. The elevated water levels and large waves accompanying the passage of a hurricane across the Gulf of Mexico are examples of short-term departures from the average physical conditions in the Gulf that impose abrupt changes on the form of the coastal features.

Also the large-scale structures built at Half Moon Bay changed the physical environment within the newly protected area outside it as well. This led to new patterns of erosion. The present coastal changes represent a tendency toward a new state of equilibrium.

Those who must deal with such changes or who must make decisions concerning the use of coastal land need to know what effects can be expected, given specific changes in the physical and biological environment. The major variables are: (1) mean water level, (2) wave climate, (3) current patterns, (4) sediment supply, and (5) properties of the coastal materials. These variables are discussed in the following sections.

Water Levels

The specific location of the shoreline at any moment is determined by the water-level elevation at that moment, and the average location of the shoreline during any time interval is determined by the average water level during that time interval. This is an obvious but important point, which is illustrated here for the Great Lakes and the oceanic coasts.

We have seen that water-level elevation on any one of the Great Lakes displays an annual cycle of variation controlled by seasonal variation in precipitation and evaporation and by long-term fluctuations related to long-term variations in precipitation over the entire Great Lakes system.

In 1932, when water levels on the Great Lakes were at record lows, the shoreline moved far out from its long-term average position. In 1972, when water levels were close to record highs, the shoreline moved landward.

During periods of high water, bluffs that had been safe from wave attack were suddenly confronted by breaking waves, which caused erosion. Newspaper reports on this kind of shoreline erosion are common during such periods of high water; few, if any, such reports appear when lake levels are low.

The location and form of many features along ocean coasts are controlled by the elevation of the mean water level along those coasts. This mean water level lies between the mean high and mean low water levels (MHW and MLW) where tides are a factor.

The elevation of these mean levels changes over time, as explained below, and the elevation of the land surface also changes due to subsidence or uplift. These changes in the relative positions of the mean water level and the land-surface elevation determine the position of the shoreline and the locus of the attack of waves on the land.

This may be seen from a reexamination of Figure elsewhere in this chapter, which shows stages in the migration of a barrier island. The migration is controlled by rising sea level. The sea level rose to its present position when melting of continental ice sheets added water to the oceans. Tide-gauge records show that the sea level is still rising and has risen 12 cm in the past century.

These authors have studied this trend and believe that the rise will continue; they predict that the sea level will rise 20 to 30 cm by the year 2050 and maybe as much as 60 cm.

The recent increases are believed to be related to the warming of the oceans. As cold ocean water is warmed under the influence of increasing mean annual atmospheric temperature, the volume of ocean water is increased. The effect of such an expansion is shown in Figure elsewhere in this chapter by the dashed and dotted lines.

If the rate of melting of glacier ice in the Arctic, the Antarctic, and Greenland also increases, then an influx of water from the melting glaciers would increase the volume of ocean water. This would then contribute to a rise in sea level. A steady rise of this sort will make itself felt by controlling the patterns of erosion and inundation in the coastal zones throughout the world.

Waves

The chief agent responsible for the incessant movement of material along a beach and for the erosion of beaches and bluffs is *wave energy*. A complete review of the formation and importance of waves is to be found in Silvester. This review has provided the standard factual material cited here. Waves are generated by the wind as it moves across open water; the energy is in the wave form.

The movement of these waves shoreward provides a means for transferring energy from the wave form to the water itself in the shore zone, where energy in the breaking waves can be expended in the work of eroding and shaping the coast.

The types of waves and the distribution of wave types differ from one coastal segment to another, so each coastal segment is influenced by a particular wave climate, just as land in the interior is influenced by a particular meteorologic climate.

The main features of waves are illustrated in Figur elsewhere in this chapter. The periodic rise and fall of any part of the wave is accomplished by the orbital (circular) motion of water particles as shown. This orbital motion, as measured by the diameter of small circles, is greatest at the surface and decreases downwards, becoming negligible at a depth (D) equal to one-half the wave length.

There is little net movement of the water particles. Wave steepness is the ratio of wave height to wave length. Waves for which the ratio of height to wavelength is relatively large are described as *steep* waves. When wave steepness reaches values between 1/7 (deep water) and 1/10 (shallow water), the wave becomes unstable and *breaks*.

Two types of waves are generated by the motion of wind over water. Waves generated by storm activity on the open ocean are called *a sea*. These waves are steep and they display a variety of periods and wave lengths.

These storm-generated waves form in an area of the ocean called the *fetch*. The waves that make up a sea are gradually reorganised in the open ocean into a set of waves that are much more regular and much less steep and are characterised by periods measured in seconds. These highly regular waves are called *swell*.

These more regular waves can travel long distances across the open ocean before impinging on coastlines. As a result, different coastal segments are shaped under the influence of the swell that is characteristic for a particular coastal segment. This swell contributes to a particular *wave climate* for a particular stretch of coast.

In deep water where water depth is greater than half the wave length, the waves move landward at a velocity or *celerity (C)*, governed by the equation C = L/T. In deep water, the waves do not interact with the bottom, and they are therefore not slowed by the frictional effects of the bottom.

As the waves approach the shore, the orbital movement is affected by the bottom. As water depth approaches a value of L/2, the wave begins to interact with the bottom, and this interaction leads to a slowing of the incoming wave. This interaction between wave and bottom leads not only to this decrease in velocity but also to an increase in wave height (H). As the waves steepen, they begin to break.

Breaking occurs as the waves enter the zone in which the water depth is 1.3 times the wave height. The zone in which the waves break is called the *surf zone*. Within this zone, part of the energy carried by the waves is expended in moving material on the bottom of the surf zone.

The fate of this material depends on the nature of the incoming waves and on the current patterns set up in the shore zone by these waves. Under the influence of incoming swell, the net movement of bottom material is shoreward, and the shore zone remains stable.

Under the influence of steeper storm waves, such as those experienced during the winter on many coasts, the breaking waves not only scour material from the shore zone but also ensure that this material is transported seaward.

Currents

As a wave approaches the shore and is influenced by the gradually shoaling conditions, the wave is slowed. Only the first segment of a straight wave crest to arrive in shallow water is slowed; the rest of the wave crest continues to move at its original velocity. As a result, the originally straight wave crest is bent.

This bending is called *wave refraction.* This refraction pattern may be observed readily from the air, and it may be represented graphically by a wave-refraction diagram. This diagram, prepared for Half Moon Bay, shows a series of straight wave crests oriented N-S.

These are the crests of incoming swell. The diagram also shows a set of straight lines orthogonal to these crests. Originally, each wave-crest segment between the two orthogonal lines carries the same energy as the next segment.

Note that the wave-crest lines are bent sharply as they encounter shallow water and the breakwater at the north end of Half Moon Bay. This process of wave refraction concentrates wave energy on specific segments of the coast and on structures along the coast. The energy that

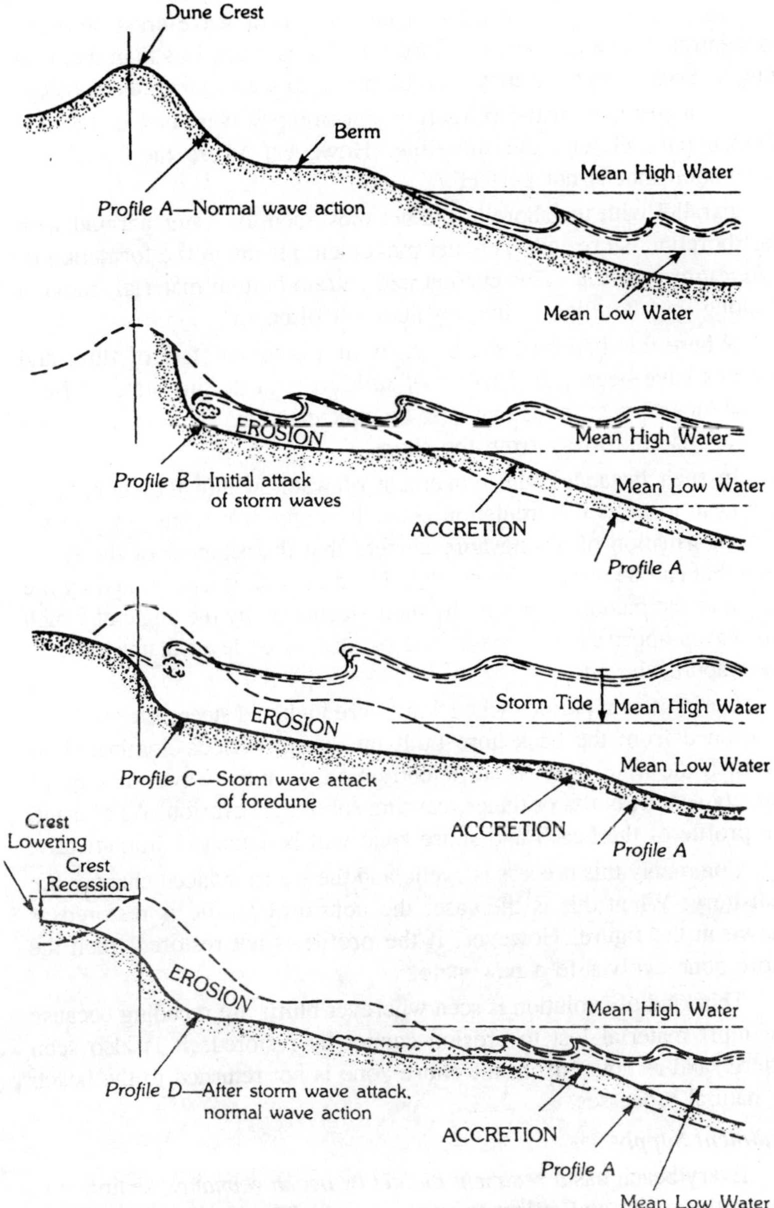

Figure 8.8 : Changes in beach profile resulting from temporarily elevated water levels.

was uniformly distributed along a unit length of wave crest becomes concentrated on a segment of shore line that is much less than that unit length. Such a segment may then be prone to wave-generated erosion.

As a result of wave refraction, incoming waves tend to become aligned parallel with the shoreline. However, where they retain an orientation that is not perfectly

parallel with the shoreline, a net movement of water parallel with the shoreline is created. This net movement results in the formation of a *longshore current*. This current can entrain bottom material, moving it along the shoreline in the downcurrent direction.

Where this happens, the beach is in a constant state of flux, and beaches have been called *rivers of sand* to indicate this state of flux. These longshore currents may also be punctuated by *rip currents,* which deliver material away from the shore.

In such instances, the movement of water within the shore zone occurs in a series of circulation cells. It is apparent from this process of the formation of a longshore current that the segment of the shore zone that can be affected by wave-induced erosion is determined by the height of the incoming waves, by their steepness, by the angle at which the waves approach the coast, and by the water level at the time of wave approach.

During storm periods when waves are high and steep, material will be eroded from the backshore built up during periods dominated by swell. If the water level is sufficiently high, then the waves may break directly against bluffs or dunes, causing substantial erosion. As a result, the profile of the beach and shore zone will be changed dramatically.

Commonly this process is cyclic and the storm-induced changes are transitory. When this is the case, the nonstorm profile is restored as shown in the figure. However, if the profile is not restored, then the shore zone evolves to a new state.

This kind of evolution is seen wherever bluffs are receding because the bluff material lost to erosion cannot be restored. It is also seen where sand removed from the shore zone is not returned to the beach by natural processes.

Sediment Supply

Every beach has *a sediment budget* or *beach economy;* sediment is fed to the beach and is lost from the beach. The nature of the beach economy is easily appreciated along the California coast, where it has been shown that many beaches are fed by sediment-laden rivers. Some fraction of this sediment passes through the beach and ultimately is lost to deep water. This loss is usually made up by steady additions from the river or other source. These segments of the coast that include a supply, a beach, and a point at which sediment moves to deep water are called *littoral cells*. Each such cell has its own sediment budget.

Beaches in settings like the Gulf Coast, the Atlantic Coast, and the Great Lakes also have their own sediment budgets. Much of the supply in these settings comes from erodible bluffs and blankets of sediment lying on the broad continental shelf or on lake margins.

These supplies are in addition to stream sediment. Sediment stored in dunes may also serve as a third source for beach development during and after storms. In these settings, little sediment is lost to deep water, but it may be stored in accumulated sand deposits such as spits and bars.

Predicting Coastal Erosion

We can and do observe dramatic changes in coastal forms, changes that occur on time scales ranging from hours to centuries. The degree of public and governmental concern with these changes is related to the time scale on which they take place.

The shorter the time scale, the greater the concern. Yet long-term coastal changes are also important, and they deserve to be examined. Such examination has been facilitated by the development of aerial photographic methods, which provide timeseries frameworks in which the findings of those who study coastal change on the ground and in the water can be placed.

These frameworks can be corrected for photographic distortion by computer-assisted processing, making it possible to monitor coastal change with a high degree of accuracy. The use of these methods in evaluating and predicting coastal change may be illustrated for barrier islands.

Barrier islands undergo incessant change under natural conditions. These changes are readily evident on the beaches facing the sea and around tidal inlets, but changes are not restricted to these locations.

The set of changes affecting these islands over time provides the means by which entire islands shift their shape and position, leading to the often astonishing changes seen from a comparison of maps or aerial photographs made at intervals of a few years.

These changes in the position of the barrier islands indicate that the elements of the islands-dunes, beaches, fringing marshes-have themselves migrated laterally.

Coastal geologists have measured the rate of migration of barrier islands in the different natural settings in which they occur and have determined the natural controls on this rate. The results of one such study showed that 12 barrier islands off the Virginia coast have migrated landward during the past years.

The northernmost group has moved landward from 247 m to 607 m in this period. The middle and southernmost islands have also retreated, shifting their orientation as they migrated. This pattern of retreat has been controlled by sea level rise, by subsidence of the floor on which the islands are built, and by wave-refraction patterns.

This particular set of islands shows the highest rate of recession observed in the 630-km stretch of coast between New Jersey and South Carolina. The average rate of recession in this segment is 1.5 m/yr. Some of the islands in this segment display accretion rather than recession; these generally are developed islands on which a variety of engineering tactics are being used to counteract the natural tendency for erosion to take place.

Comparable studies along the Gulf Coast have established that the Gulf Coast barrier islands have been moving landward or receding during the past 100 years. Wilkinson and McGowan suggest that the rate of this recession may have increased 30 to 40 percent in recent times due to human actions.

Long-term Forecasting of Barrier-island Migration

On the basis of studies such as the above, coastal geologists have developed some confidence in forecasting long-term, large-scale coastal changes in natural systems, and they and coastal engineers have alstlearned how to predict shorter-term, smaller-scale coastal changes that can be expected to follow various human actions.

Geologists have argued that long-term forecasts may be made by extrapolating historic trends. For example, if the observed rate of change described for the Virginia barrier island complex is multiplied by the length of a future time period, then the total recession expected during that time period can be estimated.

The confidence with which such an estimate can be made is affected by the uncertainty inherent in estimating the historic rate of change and by the uncertainty about the continuation ion of change in controlling variables such as sea level.

The implications of estimates of coastal recession may be seen from maps that show the observed rate of recession, the total recession in a prior time period, and the predicted position of the shoreline at the end of a future time period.

A sample of such maps shows that the shoreline of Fenwick Island, Maryland, is expected to move inland along much, but not all, of the island, encroaching on existing development, and, of equal importance, pushing the storm-surge line inland also. The shoreline at the southern end of the island is expected to move oceanward, due to accretion against the jetty built at the southern end.

Forecasts of this sort call attention to long-term trends, but a more complete local analysis is required if predictions are to be made about specific coastal segments. These predictions are referred to here as *dynamic predictions.*

Dynamic Predictions of Beach Change

More detailed dynamic predictions may be based on an understanding of the dynamics of coastal processes and on the expected changes in controlling factors such as sea level or lake level. These predictions are the basis for coastal-zone engineering and for the analysis of the expected impacts of human intervention or coastal engineering.

Coastal engineering typically is employed wherever coastal erosion is perceived as a threat to existing or planned uses of the coastal zone. Where such a threat is seen, efforts are made to examine the dynamics of the system, so that predictions can be made of the effects of various forms of coastal engineering.

In the United States, these predictions commonly are made by the U.S. Army Corps of Engineers. The aim is to provide a solution to a locally defined problem. These predictions are based on models of sediment erosion and transport.

Changes that take place in the beach-dune system are characterised by exchanges of material between the nearshore, the farshore, and the dune systems. The functioning of this system is critical in determining the stability of coastal forms. The truth of this assertion is most readily appreciated wherever sediment supply has been cut off by human action.

Sediment supply to a beach may be interrupted by damming a river or placing a structure along the shore. In the case of river damming, the cutoff of sediment supply leads to a decrease in beach size.

Therefore, if the beach is to be maintained, it must be artificially supplied with sediment. The magnitude of the original sediment input must be determined in order to estimate the scale on which the beach must be artificially fed or nourished.

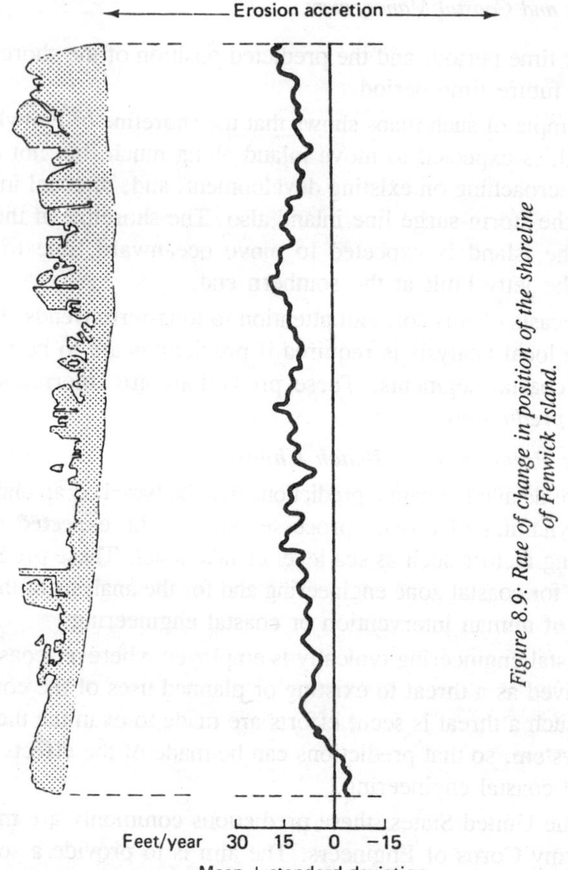

Figure 8.9: Rate of change in position of the shoreline of Fenwick Island.

This kind of budget analysis has been made for California beaches such as Silver Strand Beach south of San Diego. The Rodriguez Dam cut off the sediment supply to Silver Strand Beach, and the beach is now fed artificially. Finding a source of sediment and creating an artificial feeding system are, of course, costly problems that can be difficult to solve.

Placing structures along the shore also interrupts sediment transport so some beaches will grow and some will decline. The general results are easily estimated, but fully predicting the long-term effects of the construction of coastal structures is not so easily accomplished. We can examine situations in which the structures are placed with the express intent of trapping sediment to ensure beach growth and other situations in which the structures are built to accomplish other ends.

A groin is a concrete or rock structure constructed to interrupt the longshore flow of material and to trap sediment on the upcurrent side of the groin. This intent is easily realised. However, trapping of sediment creates a supply deficiency downcurrent of the groin, with the common result that erosion is induced downcurrent of the first groin to be placed.

In common practice, as soon as this erosion begins to occur, additional groins are placed, leading to a chain reaction of trapping and erosion along the coast. This sequence of events is illustrated in Figure elsewhere in this chapter.

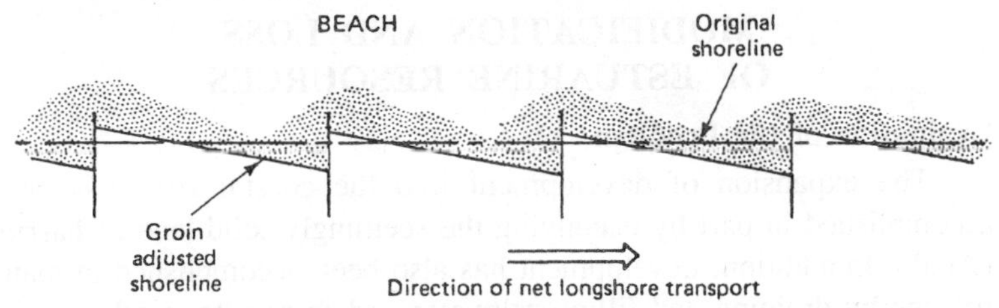

Figure 8.10 : Effects of the placement of a series of groins to intercept sand.

Although the sediment flux along the coast cannot be measured directly in most settings, it can be estimated from the rate at which erosion of a source occurs or from the rate at which accretion at a sink takes place. Along oceanic coasts this sediment flux ranges from 150,000 to 1,600,000 m^3/yr, and along the shores of the Great Lakes the range is from 1000 to 150,000 m^3/yr.

Interruption of sediment flow on this scale can lead to large effects. This is well illustrated by the relationship between the south end of Fenwick Island, which is controlled by a large groin, and the north end of Assateague Island south of this groin. Originally, these islands had shorelines that lay along a single straight line. Now the shoreline of Assateague is offset westward because its northern source of sediment has been cut off, causing erosion and landward movement of Assateague.

Jetties and *breakwaters* are structures built to provide harbors that are protected from waves. These structures inevitably have the same effect on sediment flux as groins. This is illustrated at Half Moon Bay, where breakwaters at the north end of the bay provided a protected harbor.

The breakwater had the unintended, and perhaps unexpected, effect of modifying wave-refraction and longshore-current patterns. As a result, coastal erosion and deposition patterns were also affected. Erosion was accelerated south of the breakwaters because of the breakwaters.

In each of these situations, and in many others as well, it is perhaps fair to say that the immediate local effects of the construction of a groin or a breakwater were correctly predicted.

However, it is not at all clear in many instances that efforts were made to predict the long-term and areally more extensive effects of sedimentsupply interruption. Thus we must view the use of predictions in a larger context of planning and policy.

MODIFICATION AND LOSS OF ESTUARINE RESOURCES

Evaluation of the Problem

The expansion of development into the coastal zone has been accomplished in part by occupying the seemingly solid land of barrier islands. In addition, development has also been accomplished in many regions by draining and filling saltwater and freshwater wetlands and filling parts of estuaries.

This has led to wetland losses. Simultaneously, the occupation of all these coastal areas has led to increased contamination of many estuaries with sewage effluent and with organic chemicals. Such actions have led to a loss of resources.

These losses have engendered three principal concerns: (1) the decrease in carrying capacity or productivity of ecosystems, (2) the placing at risk of a growing population, and (3) the loss of aesthetically valued regions. We have dealt with the second of these in discussing barrier-island policy and also in considering flood hazard. Therefore we will consider the first and third jointly.

In an earlier era, modifications of wetlands and estuaries were viewed as signals of progress, and this view certainly is still held in some quarters. However, an alternative view has taken substantial hold since World War II.

This alternative view of estuarine and wetland resources is founded principally in a concern with the loss of fish and wildlife habitat. Thus any modification of an estuary or a wetland that results in a loss of habitat will be regarded as a loss of resources by those who hold this view.

This way of looking at these resources is rooted in the work of estuarine and wetlands ecologists and geologists, and it has found public expression in the work of the U.S. Fish and Wildlife Service and in national conservation organisations such as the Conservation Foundation, the Nature Conservancy, and the Audubon Society. An overview of this position may be found in Clark, and in surveys by the U.S. Fish and Wildlife Service.

The U.S. Fish and Wildlife Service has estimated that there are 15 million acres of coastal estuary in the United States, about half of which it views as important habitat. Between 1947 and 1967, the Fish and Wildlife Service estimates a loss of 8 percent of this habitat. This national estimate conceals extraordinary differences.

For example, during that 20-year period, California lost 67 percent of its estuarine habitat, while Maryland lost only 0.3 percent. Similar attempts have been made to estimate the loss of coastal wetlands. The U.S. Fish and Wildlife Service estimates that less than half of the 11,000,000 acres of coastal wetlands present in 1780 still remains.

The Service also estimates the *annual* loss of coastal wetlands in the period between 1954 and 1978 to be about 100,000 acres. Let us sample some more detailed views of these losses.

Monitoring Biological Change-molluscs

One way to examine biological change is to monitor the distribution and relative abundances of various molluscan species. These are very useful biological indicators for several reasons. First of all, they occupy well-defined zones in mud and sand substrates of estuaries, and these zones can be mapped.

Also, molluscs such as clams, oysters, and mussels have been commercially important for a long time, and good records are available on harvests. In addition, they are bioaccumulators that absorb contaminants from the water creating higher concentrations of some contaminants in the organism than in the water.

The use of molluscs as indicator species has been reviewed by the Council on Environmental Quality (CEQ) in a report that provides a view of one kind of estuarine change and loss. This report is the source for the following overview.

Most studies of the size of commercial harvests of shellfish taken from estuarine waters of the United States indicate that these harvests have been declining. In some instances the declines have been in progress for many decades; in others they apparently date from the 1960s on.

For this reason, it is evident that declines cannot be attributed to one single factor. The CEQ has concluded that four effects of human action have contributed: (1) overfishing, (2) loss of habitat, (3) changes in salinity of estuarine waters, and (4) pollution.

There is a consensus that overfishing has contributed to the declines, but there is also considerable agreement that the other three factors, which are determined by management approaches to land and water, are also important. Let us look at each of these three.

1. Habitat loss is caused by dredging and by filling of estuaries. The U.S. Fish and Wildlife Service estimates that 4 percent of important habitat was lost through dredging and filling in the period 1950-1969. The greatest losses were in San Francisco Bay.
2. Changes in salinity can be caused by modifications of the existing influx of fresh water. For example, the introduction of large-scale municipal sewage treatment plants as proposed for eastern Long Island would route water that has previously returned to the ground through septic tanks directly to ocean discharge points.

   This would decrease the ground-water flow to the bays on the south side of Long Island. In studies of the possible effects of such diversion, the U.S. Environmental Protection Agency reported that salinity of Great South Bay had already increased from 23 parts per thousand in 1930 to 26 parts per thousand or higher in 1980.

   Since the spawning and survival of larvae of hard clams are most successful in waters having a salinity at the lower end of this range, any further increase would be expected to diminish productivity of these species.
3. Contamination of two types is well-documented. Contamination of shellfish waters with sewage effluent is monitored by public health departments, and these records have been summarised by the U.S. Environmental Protection Agency.

   This type of contamination results in significant resource losses because these shellfish are banned from human consumption. In addition, contamination by pesticides, PCBs, and heavy metals has also been monitored. One of these monitoring programs, called Mussel Watch, begun in 1976, has shown significant contamination by PCBs as is particularly evident along the southern New England coast.

Filling and Conversion of Wetlands

In the years since World War II, the straightforward filling of coastal wetlands has converted tens of thousands of acres to solid land used for urban development and industry. This kind of conversion is monitored by comparing aerial photographs taken at different times. Lins employed this technique to demonstrate that during the period 1945-1975, this process had removed 80,000 acres of barrier-island wetlands from their function as habitat.

The National Oceanic and Atmospheric Administration (NOAA) estimates that current annual losses from all types of coastal wetlands are on the order of 200,000 acres. The filling and paving over of wetlands also cause clear and irretrievable losses of habitat. This is not the only kind of conversion.

Along the Atlantic and Gulf coasts a technique called "finger fill" development has totally transformed coastal wetlands and estuaries. In this technique, the wetlands are filled and cut into rectilinear geometric patterns, and the estuaries are deepened to form sets of parallel channels. The full effect of this change in wetlands and in water circulation and water quality is not yet known.

MANAGING COASTAL RESOURCES PLANNING AND POLICY

The pressure to use the coastal zone for a variety of purposes is intense, and we have every reason to believe that this pressure will not diminish. Studies of the conflicts among coastal zone users document this; so too do quantitative studies of changes in coastal land use. Maps of these land-use changes show how rapidly use of the U.S. coastal zone has intensified.

As more and more users crowd closer and closer to the shore, it is inevitable that the incidence of coastal problems will increase. The problems may be described as if they are separate and distinct-erosion problems, flood problems, and so on.

However, the limited success met in dealing in isolated fashion with some of them suggests that some form of unified management is desirable. Here we will look at management of one type of problem-erosion-and then show how this might be dealt with in a more comprehensive attempt to deal with the whole set of coastal issues.

Historical Adaptations to Coastal Erosion

Adaptations to coastal erosion other than simple avoidance include: (1) structural control of erosion, (2) control of sediment flow through the use of devices that restore elements of the natural system, (3) control of sediment exchanges through artificially imposed changes in vegetation, (4) distribution of losses through insurance, and (5) control and modification of market-driven land-use practices. Few adaptations to coastal erosion can be effective on a small scale.

An individual can choose to avoid the coast, but an individual cannot control the rise of sea or lake level. Because of this, most adaptations to coastal erosion depend on actions taken by institutions, not individuals. This may be appreciated by tracing the development of policy that bears upon coastal erosion in the United States and by tracing the history of efforts to find new ways to adapt to old problems.

State and federal policies for adapting to coastal erosion problems date from the 1920s and 1930s, paralleling the development of soil erosion policies by the federal government. Evolution of policy from that time reflects a monotonic growth in scientific understanding and an improved ability to look at coastal erosion problems on larger and larger scales.

The pioneering policies for dealing with coastal erosion gave the U.S. Army Corps of Engineers responsibility in the 1930s for employing structural solutions to erosion problems and for carrying out research at what is now the Coastal Engineering Research Center (CERC). Between 1930 and the end of World War II in 1945, most erosion problems were dealt with by the construction of sea walls, jetties, and groins.

Large sea walls were constructed during the 1930s along substantial segments of the coast of the northeastern United States. Many of these sea walls remain in place. These were built by the U.S. Army Corps of Engineers and other federal agencies.

These structural solutions were costly. Experience demonstrated that they had to be implemented on a large scale if they were to last for more than a few years. In addition, experience with the unintended effects of some of these structures, such as cutting off of sediment supply, led to consideration of alternative policies.

Improved understanding of processes of longshore transport of sediment and of the sediment budget of littoral cells led to policies calting 6 !he use Of AMstructural engineering techniques. In the 1950s, the technique of artificially nourishing beaches came into fairly widespread use, and, then, in the 1960s, the additional technique of using systems-called *bypass systems*—that would maintain sediment flow through or past structural barriers was adopted.

Experience with these beach-nourishment systems and with bypass systems showed that they could be effective and that they could bring beach erosion under control. This experience also demonstrated that these systems required large and continuing expenditures of money. Furthermore, they required supplies of sediment. Providing these was not easy. The use of sources such as sand dunes or the nearshore region, while successful in the short run, interfered with the functioning of the larger beach economy under the stress of storms.

If sand was taken from the dune system to provide for beach nourishment, the dune system was not available as a sediment source during storm erosion. Similarly, if sand was drawn from the nearshore region, the entire shore was unintentionally reshaped, and a new profile evolved.

These nonstructural techniques, introduced in the 1950s and 1960s, had a precedent in efforts made in the 1930s to stabilize dune systems on barrier islands by planting and force-feeding plants such as sea oats. These plant species were introduced and thrived in the dune environment.

They trapped windblown sand more effectively than the original plant species. As a result, the dunes grew vertically and seemed to become more stable. These experiments, carried out by the National Park Service, have been analysed by Godfrey and Godfrey and by Dolan et al..

These analyses showed that these seemingly benign tactics, which mimicked natural systems, changed not only the barrier-island profile but also the distribution of vegetation, the response of the beach-dune system to storm attack, and the delivery of sediment to the back-island lagoon.

Storm-induced erosion of the artificially stabilised beach-dune system was more severe and long-lasting than erosion of the natural system. This may be seen from a comparison of profiles of the beach and dune systems eroded by Hurricane Ginger.

The steepened dune front was attacked directly by storm waves, and the beach in front of the dunes was narrowed. The profile of the natural system was quickly restored after Ginger, but the profile of the artificially stabilised system was not.

We now see a concerted attack against the further adoption of structural solutions and against continued increase in use of coastal land. It is appropriate to examine this new analysis.

Two Views of Coastal Land-Use Management

A reexamination of any section of this book will show that one can almost always identify two seemingly opposed views about the use of land and water resources. One view is that resources are there to be developed by anyone able to use them; the other view is that resource development should sometimes operate under constraints suggested by scientific or economic analysis. We saw this in the early conflict between the U.S. Congress, which wanted to develop the West as quickly as possible, and J. W. Powell, the second director of the USGS, who urged the use of scientific analysis of water resources in guiding development.

Luna Leopold recognised that this pairing of views is a recurring phenomenon, and as President of the Geological Society of America, he addressed the membership, observing that this pairing is well represented among geologists who see the same situations from different perspectives. He wrote in a letter to the GSA membership in October 1972:

In fact, it is apparent to me that a schism exists within the geological community. One group emphasizes the need for more exploration and faster development of earth resources to meet the rising demands of a technological society. This group continues the philosophy of the economic geologist of past decades whose primary concern was the free play of competition in the discovery, the exploitation, and the production of mineral and earth resources. . . . Contrariwise, there is a small group of geologists among those labeled conservationists or environmentalists, who are accused by some of their colleagues as being in favor of locking up resources. . . .

The debates about the coastal zone illustrate these alternative ways of viewing familiar situations. It is appropriate in these concluding sections to examine contrasting views on coastal erosion and the application of scientific knowledge in guiding coastal development.

Two Views of Coastal Erosion

Many who have occupied the coastal zone have been or are now being threatened by erosion and flooding. They have sought and obtained structural protection. Sea walls and groins have been built, and beaches have been nourished. Land has been made more valuable, and valuable land has been protected.

Those who have urged these actions or implemented them have given first priority to development. The coastal geologists at the Conference on America's Eroding Shoreline argued forcefully against this approach. They urged an end to most, not all, beach stabilisation, the adoption of regulations to create an undeveloped coastal buffer zone, and restrictions on the use of public funds for poststorm redevelopment. In other words, they took a conservation position and a position opposing expenditure of public funds for private benefit.

They supported this position by arguing that the public, the policy-makers, and the builders of coastal structures are not aware that the sea level is rising and do not realize the implications of this rise. They argued that these people must be made aware through consideration of sources such as Pilkey et al. (1983).

Two Views of Geology in the Service of Coastal Development

Just how sharply geologists can differ about familiar coastal situations was illustrated in 1975 and 1976 when two geologists published a short paper on the presentation of coastal information to real-estate brokers and builders. In this paper, Mathewson and Piper showed how geologic information about Padre island on the Texas Gulf Coast could be translated into terms that showed the level of expenditure required to make use of land on this island for development.

This approach may be sensed from the concluding paragraph of their paper:

In the study area selected, the low-value beach and foredune land would be well suited for high-rise, high-cost hotel and commercial structures that can absorb the higher costs necessary for protection. On the other hand, the dune-field land would be suited for high-rise structures and high-density housing, thus making it economic to stabilize or remove the entire field.... In this example, the entire 12.59-km^2 tract of land could be developed; however, specific land uses were based on the geologic factors of the site, thus increasing potential profits and providing consumer protection at the lowest percentage cost of each structure.

A year later, two commentaries on the Mathewson and Piper paper were published, one by R. S. Kerr, and the other by C. M. Woodruff, W. L. Longley, and A. Reed. These appeared in the May 1976 issue of *Geology,* and they were followed by a reply from Mathewson and Piper.

These commentators sharply criticised the views of Mathewson and Piper, arguing, for example, that their conclusions were "totally in opposition to the conclusion dictated by studying the geologic processes active on Padre Island." Piper, in return, argued that geologists evaluating the geologic aspects of land-use planning have the added responsibility to see that their judgments, often accepted as fact, are based on sound geologic and ecologic principles. In the coastal zone, in particular, ill-advised development can lead to serious loss of both life and property.

This reply by Piper seems to support the arguments advanced by critics of Mathewson and Piper.

This dialogue deserves to be examined in the original paper and the commentaries because it shows with great clarity that geologists probably equally familiar with a particular coastal setting (Padre Island) can hold very different views about the appropriate use of geologic information and development. Anyone interested in doing so might prepare by reviewing the appropriate section of the *Environmental Geologic Atlas of the Texas Coastal Zone,* published in sections by the Bureau of Economic Geology in Austin, Texas.

We have considered some rather specific responses to a specific type of problem. We have seen how changing scientific views combined with economic analysis of costs can lead to new tactics. Now let us look at this in a more general contextthe comprehensive management of coastal resources.

STATE AND FEDERAL POLICY ON COASTAL RESOURCE MANAGEMENT

In 1972, the U.S. Congress passed the Coastal Zone Management Act of 1972, in recognition of the importance of issues that have been described in this chapter. States were to be given funds to allow them to engage in the planning of a coastal zone management system.

If their proposed system was approved, they were to be given additional funds to aid in implementing the system. Thirty-five states and 211 U.S. Territories became eligible for assistance from this program. By 1981, 21 of the 35 had approved programs. On 17 October 1981, the Coastal Zone Management Program was authorised for another five years.

At the end of the first five years of the program, the Office of Coastal Zone Management assessed progress, and it named four main areas with which it thought states should continue to be concerned:

1. Protection of significant coastal resources and areas, such as wetlands, fisheries, beaches, dunes, and barrier islands.

2. Management of coastal development to minimize loss of life and property because of improper development in areas subject to coastal hazards, giving priority to coastal-dependent development including energy facility siting and identifying sites for dredge-spoil disposal.

3. Increased access to the coast for recreational purposes, including revitalisation of urban waterfronts and protection and restoration of cultural, historic, and aesthetic coastal resources.

4. Improved predictability and efficiency in public decision making, including increased intergovernmental cooperation and coordination.

Let us look at the first two goals listed above. One is concerned with the protection of resources, the other with the protection of people and property. If these are desirable goals, then any coastal zone management program might attempt to encourage development of policies that meet both goals, policies that will not act in opposition to one another.

Such a program might also guide economic and scientific analysis. Let us look at some general principles and approaches first and then examine the experience of a single state-Rhode Island.

There are two types of coastal environments in which the issue of resource protection and the issue of the protection of people are invariably confronted. These are coastal wetlands and barrier beaches. Any proposal to place housing on a wetland in a hurricane-prone area is a proposal to give up wetland resources and to place people and structures at risk.

The same argument may be applied to the development of the ecologically and geologically most vulnerable parts of barrier islands. Most analysts understand this argument. Yet many policies that partially regulate land use in these two environments have not been as successful as hoped. The same may be said for some technical and economic analyses that have sought to providc a rational basis for regulation or nonregulation.

We can see this from sampling efforts to regulate barrier beaches and coastal wetlands. With the passage of the National Flood Insurance Program in 1973, a form of barrier-beach development regulation seemed to have been created. New development backed by federal funds came under the rules of the National Flood Insurance Program.

Since the insurance rates were low, however, this program actually encouraged development. Homes were built in vulnerable locations and were then insured. During events such as Hurricane Frederic, these same homes were destroyed. Some were destroyed because of erosion rather than flooding, but many owners of such structures were compensated for their losses. And they prepared to build again.

Economic analysis of these losses convinced many in the federal government that the insurance program should be changed-and it was. Rates were increased. During this same period, the coastal scientific community was effectively making its case that many of these barrier-island settings were highly vulnerable in the long run, and they were also making the case that modification was causing ecosystem modifications on and adjacent to the barrier islands.

As a result, the various actions cited above were taken. Let us look more carefully at this because it illustrates the development of a more integrated view of management. The influence of this point of view and of the experience with the National Flood Insurance Program has led to three interesting decisions that collectively illustrate what could become facets of a larger-scale approach to coastal management.

The first decision was taken by the U.S. National Park Service in 1973, when it decided that it would no longer attempt to stabilize barrier islands. The second decision was taken by the U.S. Congress when it passed the Barrier Island Resources and Development Act in 1982. This banned the further use of federal subsidies to provide bridges and protective structures for undeveloped islands.

The third decision was the passage of legislation in 1981 banning federal flood insurance for new development on specified barrier beaches under the National Flood Insurance Program.

The Coastal Barrier Resources Act of 1982 prohibited any federal expenditures on a set of barrier islands shown on maps provided with the legislation. This legislation got extraordinarily broad support in the Congress even though it was opposed by the National Association of Realtors and the National Association of Home Builders.

It got this support because it simultaneously dealt with the two major issues-protection of resources and protection of people. Recognition of this was expressed by Representative Evans, sponsor of the House version of the bill, when he remarked that "very rarely, if ever, do Democrats and Republicans have an opportunity to vote for a bill that saves lives, saves money, and helps protect our environment."

This kind of recognition has not yet been given to the wetlands situation as may be seen by considering the U.S. Army Corps of Engineers policy on wetland development permits and by considering the views of several economists.

The U.S. Army Corps of Engineers has permit authority over coastal wetlands. Before the District Engineer can issue a permit, he must determine that the benefits of the proposed alteration outweigh the damage to the wetlands resource. Note that the District Engineer must consider only the "services" provided by the wetland and need not consider the hazard.

This focus on services is to be found in the analyses of some economists who have attempted to provide a rationale for making cost-benefit analyses of wetland regulation. Let us look at an example.

Shabman and Batie note that arguments in favor of protecting wetlands typically cite the services provided by the wetlands. The Corps of Engineers' permit statement is an excellent example. Shabman and Batie chose therefore to analyze the alleged benefits from these services in each of the following categories: (1) flood control, (2) erosion control, (3) food supply for shellfish, (4) waterfowl habitat, and (5) removal of contaminants from water. They conclude their analysis with the statement that

> the technical linkages between wetland and natural services are not well established (so) there should be additional research focused on alternative development values of wetlands....

These authors argue that there is little sound evidence to support any of the above arguments in defense of wetland regulation. In doing so, they, too, fail to consider both the protection of the resource and the protection of the people.

They decide that wetlands do not provide much flood control. In other words, they look only at the service that the wetland is alleged to provide, and they fail to note that placing a structure on a wetland whose existence depends on its position at mean high tide is to place a structure at risk. This is amply demonstrated along the Florida coast.

Let us conclude this entire examination of coastal resources by looking at management as practiced in the state of Rhode Island. Anyone interested in another coastal state can turn to White et al. for an excellent state-by-state view of the hazard element and to the NOAA biennial reports for summaries of management practices of each state.

The Rhode Island coast contains all of the major elements discussed in this chapter. Narragansett Bay is a major estuary and an important region for shellfish. Wetlands are important in many protected areas along the bay and in the barrierbeach region along the Atlantic Coast. Contamination of estuarine resources became a significant problem after World War II.

Hurricanes have caused great damage and loss of life in the 40-year period 1938-1978. Hurricanes included the great 1938 hurricane (unnamed) and four succeeding named hurricanes. In 1971, the Rhode Island State Legislature passed the Rhode Island Coastal Zone Management Act of 1971, which established a Rhode Island Coastal Resources Management Council with authority over an extensive coastal zone.

Thus the Rhode Island legislation actually preceded the federal legislation and was one of the sources of experience to which the Federal Coastal Zone Management Office could turn.

The Rhode Island Council recognised that control of undeveloped beaches and wetlands could simultaneously protect the estuary against loss of resources and the people against loss of life and property. Thus it employs a permit system that requires that any wetland development be approved only after review.

The Rhode Island Council has been subjected to considerable pressure to allow development, and it has learned that federal programs do not always support one another. Specifically, Rhode Island, like all other coastal states, has found that the National Flood Insurance Program made it economically feasible for people to occupy floodhazard zones along the coast because they could obtain low-cost subsidised insurance that would protect them against loss of property, if not against loss of life.

Rhode Island also has a segment of barrier-beach coastline that displays all of the problems so extensively reported on further south. The University of Rhode Island Coastal Resources Center maintains erosion records for all beaches. These show that East Beach in Charlestown receded 75 feet in a 13-year period ending in 1975.

The Rhode Island Council has recognised that development along a segment of rapidly shifting coast is inadvisable, and it has exercised control over this area, control that is now supplemented by changes in the National Flood Insurance Program as we have noted.

The Council has established one of the 12 national estuarine sanctuaries in Narragansett Bay, gradually restoring the ecosystem around Patience, Prudence, and Hope islands to its earlier condition. And it has successfully controlled contamination of estuarine waters in Narragansett Bay so that many shellfish beds have been reopened to commercial fisherman.

The additional value of these areas is estimated to be $10,000,000. Finally, the Coastal Resources Management Council is developing a program for the management of the salt ponds in the coastal zone.

INDEX

D

E

F

G

N

O

P

Q

R

S

T

U

V

X

Y